THE
DAWN
OF
MIND

HOW MATTER BECAME
CONSCIOUS AND ALIVE

JAMES COOKE, Ph.D.

Prometheus Books

Essex, Connecticut

Prometheus Books

An imprint of The Globe Pequot Publishing Group, Inc.
64 South Main Street
Essex, CT 06426
www.globepequot.com

Distributed by NATIONAL BOOK NETWORK

British Library Cataloguing in Publication Information Available

Library of Congress Cataloging-in-Publication Data

Names: Cooke, James, 1989– author.
Title: The dawn of mind : how matter became conscious and alive / James Cooke.
Description: Lanham, MD : Prometheus, [2024] | Includes bibliographical references. |
    Summary: "Weaving together cutting-edge science and the contemplative
    insights that arise from mystical experience, Dr. James Cooke radically redraws
    our understanding of consciousness and what it truly means to be who we are"
    —Provided by publisher.
Identifiers: LCCN 2024013779 (print) | LCCN 2024013780 (ebook) |
    ISBN 9781633889927 (cloth) | ISBN 9781633889934 (ebook)
Subjects: LCSH: Consciousness. | Thought and thinking. | Self-consciousness
    (Awareness)
Classification: LCC BF311 .C673 2025 (print) | LCC BF311 (ebook) |
    DDC 153—dc23/eng/20240517
LC record available at https://lccn.loc.gov/2024013779
LC ebook record available at https://lccn.loc.gov/2024013780

For Rebecca and Leo,
who illuminate my world

CONTENTS

CONTENTS

INTRODUCTION

What Are You?

What are you? According to science, you are a collection of material particles, a complex machine, completely devoid of any inherent meaning. According to certain religions, you are a divine, conscious soul with the capacity to feel and love, imbued with meaning and cosmic significance. How did we arrive at such seemingly opposite conclusions on what it is to be a human being? The answer lies, in part, in the fact that a human being can be understood from two perspectives: the inner and the outer. From the inside, it feels a certain way to be you. Whatever you might be physically, mentally you are a conscious, feeling thing. You smell the scent of the earth after it rains; you see the glory of a sunrise; you experience feelings of great variety and depth. This is the way of knowing ourselves through direct experience, the way that mystics have explored for millennia, long before we had scientific instruments with which to study the mind. Sitting in meditation, lost in ecstatic dance, or fasting and praying for days, these explorers of inner space received profound insights into our nature and our relationship to the rest of reality.

Looking from the outside, science tells us we are physical bodies made of molecules that themselves are made of atoms, which in turn are made of subatomic particles. The material substance that forms our bodies obeys laws of physics that cannot be broken, seeming to make us clockwork automatons that merely act out the desires programmed into us by evolution. It may be awe inspiring that many of the elements of our bodies were cooked up in exploding stars or that we exist through the combined riot of billions of subatomic particles, but science has no way of accounting for this feeling of awe; consciousness and feeling do not fit into our current scientific picture of the world. The scientific method has done an incredible job of helping us to understand the outer world, but the very existence of your inner world remains a mystery to science.

When we speak of consciousness, we are speaking of the very fact of experience. Consciousness is not the voice in your head or your ability to be self-aware, although these do depend on consciousness. It is the capacity to be aware of anything at all. Imagine the taste of a ripe strawberry, the warmth of sunlight on your skin, or deep pangs of longing. These experiences, so vivid and intimate, are all manifestations of consciousness. Consciousness does not require thought or the ability to self-reflect; where there is feeling of any kind, from the experience of seeing the blue of the sky to imagining what your life will be like a decade from now, there is consciousness.

An influential definition of *consciousness* comes from philosopher Thomas Nagel, who suggests that we consider something to be conscious if there is "something it is like" to be that thing.[1] For example, if I rearranged your atoms to change you into a rock, then we might presume that there would be "nothing it is like" to be the rock, which is another way of saying that rocks are not conscious. As I mixed up your atoms to turn you into a mineral object, at some point in the transition, consciousness would be extinguished, and the light of experience would go out. What about an ape or a squirrel, a termite or a bacterium? Is there "something it is like" to be them? That is, are they conscious? The challenge science faces is in understanding how our objective description of the arrangement of such physical phenomena as atoms connects to the subjective qualitative experience of consciousness, of it being *like* something to be that thing—of experiential feeling itself.

We can think of consciousness as being like a simulation. When you have the experience of seeing a rainbow, the rainbow exists in your simulation of the world, your consciousness, not in the physical world itself. We know there is no colorful arched physical structure in the objective description of the world in such moments. You experience certain things in your simulation of the world around you, and I experience something different. To understand consciousness, we must explain why you and I simulate the world around us and what leads to the content of our simulations being different. Computers are capable of simulating things, however, yet we do not necessarily think that this makes them conscious, which takes us to the core of the mystery of consciousness. Why isn't our simulation just a physical procedure that occurs in the dark? Why do our brains not simply function like a computer and physically process the necessary information without the added experience? A computer can analyze the same visual signals that give you the conscious experience of seeing a rainbow with no need for the experience. Why, in our case, are the lights on? Why is your simulation illuminated, and what is the source of this illumination?

This quality of illumination is known as awareness, and it lies at the core of consciousness. Another way to think about awareness is as knowingness.[2] When there is an experience of the taste of banana, there is the knowing of that taste. When you perform an act unconsciously, there is no experiential knowing of that act having occurred. To fully explain this inner simulation that we call consciousness, we must account for this mysterious quality of awareness that results in the simulation actually being experienced rather than merely operating in the dark through a blind, unconscious mechanism.

Though there is no shortage of speculation regarding the origin and nature of consciousness, there is no consensus on where consciousness fits into the modern scientific story of the natural world. In fact, we have no generally accepted explanation of why consciousness should exist at all. Is it the product of complex animal brains like our own? Is it the fundamental nature of our reality? Could it be an illusion? Every proposal in this array of mutually contradictory positions is held by multiple prominent philosophers and scientists today. To say there's no consensus on the issue of consciousness is an understatement.

The issue of consciousness is consequential for understanding not only our own minds but also reality itself. Reckoning with the nature of experience forces us to confront what is perhaps the most basic question that science, philosophy, and religion all try to answer: What is going on? Do we find ourselves in a clockwork universe that happened to produce some animal brains that excrete consciousness like a useless gas for no apparent reason? Are we in a matrix or some kind of simulation? Are we a dream in a cosmic mind? Taking a stance on the issue of consciousness necessarily requires us to also commit to a stance on the nature of reality. Consciousness theories and their associated worldviews are a package deal.

In this book, I lay out a way of thinking about consciousness that I previously published in the *Journal of Consciousness Studies* as the living mirror theory of consciousness.[3] The core claim is that life and consciousness are fundamentally linked—we do not experience primarily because we have brains; we experience because we are alive. The brain is certainly involved in human consciousness but as a secondary player to the life process. The evolution of the brain is not what brought consciousness into existence; the emergence of life did. The brain merely elaborates the contents of experience to admittedly dizzying heights in our species.

The living mirror theory is entirely aligned with the scientific perspective, being influenced by Darwinian theory, thermodynamics, complexity theory, biophysics, and contemporary neuroscience. As is typical when we gain scientific insight, accepting a theory comes with many interesting and sometimes counterintuitive implications. Darwinian evolution tells us that, due to the continuity of the web of life, we share an ancestor with a banana. The Copernican revolution in astrophysics led us to the realization that, rather than being at the center of the universe, "we live on an insignificant planet of a humdrum star lost in a galaxy tucked away in some forgotten corner of a universe in which there are far more galaxies than people," as astrophysicist Carl Sagan put it.[4] The living mirror theory comes with its own surprising implications for what we are.

For the theory to make sense, I must frame it within a compelling worldview. The current dominant worldview in science, in which only the material particles of physics are thought to truly exist, has left us facing a dead end when it comes to thinking about consciousness. To make sense of consciousness, we must first examine and dismantle the flawed assumptions in this predominant scientific worldview. Only then can a new understanding of our inner world and its place in reality come into focus, reconciling the scientific and religious perspectives on the question of what we are.

While the domain of spirituality and religion understandably makes some scientifically minded people uncomfortable, I believe there are valuable insights into our existential situation to be found in these traditions. In particular, there appears to be a mystical core to many religions that reflects a fundamental insight that people have found compelling throughout time: that we are not truly separate from the world around us and that we are in fact deeply at home in existence. In religious traditions, this is typically not something that one discovers by studying the world around us but instead through conscious experience itself. It seems to me that a deep understanding of this experience is an antidote to our current assumptions that block us from understanding consciousness.

I use the term *nondual naturalism* for the secular-spiritual worldview I present. *Nondual* refers to the insight that reality is not fundamentally split into subject and object, mind and matter, but is instead whole. *Naturalism* refers to a scientific perspective on reality that does not accept the existence of supernatural phenomena but instead relies on philosophy and science to map out a consistent picture of the universe, one in which the world around us contains its own explanations. The combination of the two perspectives results in a worldview in which a subset of core spiritual experiences that are in alignment with scientific findings are held to be valid. These include feeling oneself to be fully part of nature (as Darwin showed us we are) and discovering that the self is not an entity with its own independent existence; rather, one's sense of being a separate self is a psychological experience (as modern neuroscience also claims).

Finally, a note on terminology. This book deals with the central fact of our existence, that there is an experience happening. When I use such terms as *consciousness, experience, subjectivity, feeling, sentience,* or *mind,* I always refer to this simple and readily apparent fact. The feeling of touch on your skin, experiencing the scent of coffee, your perception of a beautiful landscape scene—these subjective events should not exist according to our current scientific picture of the world, and it is these phenomena that I address when I use the term *consciousness,* as well as these related terms. The term *awareness* is reserved for the core characteristic of consciousness by which the content of experience is illuminated in order to differentiate between this and the specific contents of consciousness. This all is unpacked throughout this book. I include a glossary of key terms to help keep track of the different theories and terms of art used in later chapters.

In writing this book, I am driven by more than an intellectual curiosity into the nature of consciousness. It seems to me that, at this moment, our dominant global culture is confused and lost. There is so much unnecessary suffering in the world that is caused by both our relation to ourselves and to the natural world, and it is my hope that, through greater understanding of ourselves and our place in existence, we will be able to navigate more effectively to a world with less suffering. For this reason, I have written this book to be accessible to a general audience, as well as to academics in fields relevant to consciousness. Science, when wielded wisely, can function as a light in the dark in this quest, leading us toward truth. Through greater understanding of ourselves and our situation, I hope we can move collectively in the direction of ever greater peace, both inner and outer. Even if this book fails to move us in this direction, I deeply appreciate your coming on this journey with me.

PART I

INSIDE OUT

The Philosophy of Consciousness

ONE

CONSCIOUSNESS

The View from a Dead End

Conscious experience is at once the most familiar thing in the world and the most mysterious. There is nothing we know about more directly than consciousness, but it is far from clear how to reconcile it with everything else we know.—DAVID J. CHALMERS[1]

INNER AND OUTER KNOWLEDGE

Knowledge is a powerful thing. In practically every area of life, we are more effective when armed with knowledge than when we have little or no understanding. Knowledge is arguably what humans do best. The father of modern taxonomy, Carl Linnaeus, certainly seemed to think so. When he gave our species its name, he settled on *Homo sapiens sapiens. Sapiens* comes from the Latin word for *knowledge, sapere.* We're supposedly so good at knowing that we were named for it twice. We must stop and ask, however, what kind of knowledge do we possess?

We live at a point in human history when our knowledge of the world around us has reached astounding levels. Through science and philosophy, we have come to understand not only the vast expanse of the cosmos but also the operation of subatomic particles. We know how the most majestic mountains were formed, how the elements combine to make everything from azaleas to zebras, and how the process of evolution by natural selection sculpted us over millennia into the beings we are today. Our knowledge of the natural world also gives us astounding control over it. Our technological prowess has reached a point where we can harness the power of the elemental forces of nature to run astoundingly complex computational devices like our mobile phones, devices that are more than a million times more powerful than the computers that first landed humans on the moon. Our ability to understand the world around us appears limitless.

What about knowledge of ourselves? In areas where we can treat ourselves like an object, when studying the body, for example, we have made incredible advances. We can edit individual genes so small they are invisible to the naked eye. We can diagnose and treat neurological issues occurring within the dark cavern of the skull while leaving it completely intact. Our understanding of the body as an object is as successful as our understanding of the natural world around us.

What about when we turn to subjective knowledge, to the knowledge of ourselves from within? In trying to understand the inner space that mediates our experience of life, science and philosophy fall drastically short. The contrast between our understanding and mastery of the outer world and our collective befuddlement at our inner world is truly dramatic. When we look at the very core of ourselves, at the fact that we are experiencing existence right now, scientific and philosophical consensus are nowhere to be found. The term we use to refer to this fact of experience, to the subjective, qualitative character of the mind, is *consciousness*. It is the single most significant aspect of our existence, for it is the very thing that makes the experience of significance possible. Without consciousness, nothing would matter; there would be no joy, no pain, no way for meaning to exist at all. Despite its importance, scientists and philosophers have failed to agree on what it is, why it exists, and how it relates to the world around us.

The subjective and qualitative character of consciousness can be understood by contrasting it with the objective and quantitative character of the outside world. When you dream, the experience is happening to you and no one else; it is privately occurring in inner space. It is subjective. A piece of material like a dress, though, is objective; it exists for both of us in the space between us. The shape of the dress is quantitative in that it can be described using quantities—the numbers that give you an idea of the kind of fit it might have. We might initially think that color is another objective attribute of the dress, like its shape. In reality, the physical dress only reflects light in a certain way. The qualitative experience of the color of the dress takes place subjectively, in one's mind.

The widespread assumption that colors exist out there in the world and not in here in our individual minds was put dramatically on display in 2015, when a picture surfaced of a dress that some people reported as being black and blue, while others saw it as white and gold. Many people unconsciously assumed that the dress had an objective visual appearance, leading to exasperated confusion when others disagreed about what color the dress truly was. Rather than objectively being a certain color, the dress merely appeared differently to different people; the color was a qualitative, subjective experience, not a quantitative, objective one.

If we all agree that consciousness is subjective and qualitative, then what are we scientists confused about? The issue comes down to how the quantitative and qualitative relate to each other. Galileo once proclaimed, "The Book of Nature is written in the language of mathematics."[2] He was observing that the objective natural world around us can be accurately described using mathematical models, and by starting out with this approach, the fields of physics, chemistry, biology, and the rest of the objective sciences have had astounding success. We, too, are part of nature, though, as are our qualitative, subjective experiences. How are the fields of science that deal with the mind supposed to describe in numbers the feeling of the warm sun on one's skin or account for the experience of being in love using mathematics? Where do such qualities fit into our quantitative story of nature?

This is the fundamental problem faced by those who want to truly understand consciousness. Understanding consciousness does not mean simply describing it; it involves understanding where it fits into our picture of the natural world. If we all agreed that we understood consciousness, then we would be able to tell a compelling story about where it exists and why it exists where it does, the same way we can for oxygen molecules or blades of grass. We would be able to say how and why it is that quantitative matter gives rise to qualitative experience. At this point in time, however, we are far from agreeing on such an understanding.

THE SCIENTIST'S PERSPECTIVE: BEGINNING WITH THE BRAIN

The astonishing hypothesis is that "you," your joys and your sorrows, your memories and your ambitions, your sense of identity and free will, are in fact no more than the behavior of a vast assembly of nerve cells and their associated molecules.—FRANCIS CRICK[3]

As an experimental psychology undergraduate at Oxford, I was required to participate in psychological experiments in order to receive my degree. One such experiment involved having a powerful magnet placed on my head so that its magnetic pulses could interfere with the electrical activity of my brain while I performed different tests. This method, known as transcranial magnetic stimulation (TMS), has been used to produce simple visual hallucinations called phosphenes by stimulating the neurons of the visual cortex at the back of the brain. Somehow, physical properties, such as magnetism, electricity, and the biochemistry in brain cells, come together to create a seemingly nonphysical experience, and the brain seems to be responsible for this near-miraculous capacity.

The involvement of the brain in human consciousness is crystal clear. Take a psychoactive drug, and it will alter your experience by altering the activity of brain cells. Receive an anesthetic that acts on the brain, and experience will seem to disappear altogether for the time that it is active. We do not need a modern scientific understanding of molecules and their effects on brain cells to observe the relationship between brain and consciousness. For as long as humans have existed, it has been possible to observe the effects of a head injury on the mind. This relationship wasn't readily apparent to everyone, however, as changes in other parts of the body also correlate with changes in experience. For example, alterations in heart rate vary with your conscious experience of excitement or anxiety versus feeling calm. Perhaps because of our ability to easily feel such changes in and around the heart, it was this organ and not the brain that was commonly linked to consciousness in the ancient world. While the ancient Egyptians took great care to embalm the heart of a deceased person, the brain, which they didn't even have a name for, was simply discarded. In fact, Aristotle thought that, due to its large surface area, the brain was merely a radiator that functioned to regulate the temperature of the blood.[4]

Modern neuroscientific techniques give us unprecedented access to brain activity, allowing us to probe the relationship between the brain and consciousness in ways that were impossible until a few decades ago. What's more, culture has changed to the point that such investigations are no longer considered taboo. In the decades that followed the '60s, an interest in consciousness was largely associated with hippies and New Agers. Scientists, by contrast, were interested only in things that could be objectively measured, something that is not possible in the private world of experience. Until only a few decades ago, neuroscientists wouldn't go anywhere near subjectivity.

That was until a giant of science, Francis Crick, legitimized the investigation of consciousness in the eyes of the scientific community. Crick was the codiscoverer, along with James Watson, of the structure of DNA. In his later years, along with his coinvestigator Christof Koch, he turned his attention to consciousness. Together, they embarked on an ambitious scientific attempt to account for how subjective experience relates to the objective brain by mapping the neural correlates of consciousness (NCCs).[5] This project involved mapping the core patterns of brain activity that were consistently linked to a specific experience. One way researchers did this was by showing a different image to each eye of a participant, an approach known as the binocular rivalry paradigm.[6] While the participant's brain struggled to make sense of the conflicting inputs, their subjective experience would spontaneously flip back and forth between them, though nothing had changed in the objective experimental setup. By measuring their brain activity during such an experience, researchers could see what changes tracked with this switching in consciousness while all other variables were held constant.

This was a very clever approach and a fascinating research project. It attempted to sidestep the philosophical issues that face the study of consciousness and presented itself as a purely scientific endeavor. It did not weigh in on the thorny philosophical issue of the exact nature of the relationship between the brain and consciousness; it simply mapped the correlations between them. The question remained, however, was brain activity generating consciousness? Perhaps this brain activity simply *was* consciousness? Or maybe consciousness was an illusion, a clever trick of the brain? No consensus position was reached, each stance having its own flaws. A widely accepted explanation of the relationship, one that successfully bridges the gap between the quantitative description of the material brain and qualitative experience, has yet to be found.

The scientists offered a range of potential brain mechanisms that could be responsible for consciousness, but for one person in particular, these attempts failed to address the most important question of all. In 1994, at a conference on consciousness held in Tucson, Arizona, David Chalmers—a twenty-eight-year-old Australian philosopher—took the stage for what would be an era-defining talk for the field. Here and in a landmark paper published in the *Journal of Consciousness Studies*, Chalmers laid out what he called the "hard problem" of consciousness.[7] He argued that some problems relating to consciousness were relatively easy to investigate, in that their secrets could be expected to yield to the standard scientific method. Studying how the optic nerve projects from the eye to the brain and how visual information is processed is a challenge, but there's no reason to believe that, with enough experiments, we won't ultimately come to a detailed understanding of visual processing. This is one of the many easy problems of consciousness, according to Chalmers. The hard problem is the issue of why all this processing should feel like something at all. Why should the photons from the sun that fall on my retina as our local star moves over the horizon conjure up the experience of a beautiful sunset? If our bodies, including our brains, are material systems that can be completely described in the objective, quantitative terms of science, then why doesn't all the processing in the brain just happen without experience? This is the issue we have yet to find a satisfactory solution to.

Consider the scenario of freshly baked cookies being removed from an oven. You smell the cookies and inevitably approach and eat one. What does science say is happening in this situation? Odorant molecules flow into the air when the oven is opened; some enter your nostrils, docking on specialized receptors on olfactory sensory neurons in your nasal epithelium. The activation of these receptors by the odorant molecules results in changes in the electrical activity of these cells, produced by the opening of physical pores in the cells that allow ions to flow across the cell membrane. An electrochemical relay race then occurs as a series of neurons passes on the electrical signal, temporarily transforming it to a chemical message at the junctions between different brain cells. Ultimately these electrical signals reach neurons that connect to muscles, inducing coordinated muscle contractions that result in walking toward the cookies, picking one up, and putting it in one's mouth. No matter how complex the changes that take place in the brain to mediate between this simple sensory input and this complex behavioral output, ultimately all that happens is electrical and chemical activity, physical processes that seem to have nothing to do with felt experience. In this entire story, something has been left out: the delicious smell and taste of the cookies. The presence of the conscious experience is not explained at any point in this story.

Zombies are often used to make this point—not the living-dead zombie of popular culture but the "philosophical zombie," a thought experiment introduced by Chalmers in *The Conscious Mind*.[8] A philosophical zombie is a version of a person who functions in exactly the same way as the non-zombie person but has no consciousness. Even though they have no internal experiences, they possess a body with a brain and are capable of doing everything you and I can do. They wake up and eat food; they speak with other people; their brain receives the appropriate inputs, performs the appropriate processing, and sends the appropriate outputs. All this happens with no conscious experience, however. They do not taste the food, they detect but do not hear your voice when you speak to them, they have no inner life whatsoever. Chalmers argued that the fact that some of us can conceive of such a being shows that consciousness is something entirely separate from physical mechanisms and that, as a result, no physical mechanism could ever account for consciousness. Arguably, however, it just points to the fact that we have no idea how consciousness works.

One might respond to this philosophical handwringing with the suggestion that we stop worrying about the philosophy and just turn to science. Unfortunately, the hard problem cannot be solved by simply doing more scientific experiments. We can pile quantitative description on top of quantitative description, but this alone will never get us to qualitative experience. What's more, the inherently subjective and private nature of consciousness means that it cannot be studied in the same way as the other phenomena of the natural world. You can't put the taste of coffee under a microscope or dissect the feeling of hunger. This makes the problem of consciousness fundamentally different from every other area that science studies. A complete understanding of the subjective can't just come from the scientific study of the objective; we have to take a step back and reconsider the entire way we think about experience and the world.

PHYSICALISM AND REDUCTIONISM

You might wonder if our inability to explain consciousness is really such a fundamental challenge to our existing scientific worldview. So what if there is a private world of qualitative experience that science doesn't account for? Just let science describe the rest of our universe, you might object; it can just leave my mind out of its explanations. It may seem that your private experience of seeing a sunset or smelling cookies has nothing to do with our scientific picture of the universe, but this is in fact not the case. Science provides us with a map of how the universe *behaves*; it doesn't describe what it truly *is*, apart from this behavior. As a result, something can only be said to exist in our scientific picture of the world if the way it behaves can be linked up with our understanding of the way everything else behaves. We call the stories we tell about the behavior of things *causality*. Consciousness does not currently feature in our story of causality—of how the universe behaves—so it is effectively absent from our scientific map of what exists. According to the current scientific worldview, you have never had an experience in your life.

We have a tension here between the existence of consciousness and preserving our current scientific worldview. Perhaps surprisingly, there are some who choose to give up on consciousness rather than update our current scientific perspective. Philosophers Patricia and Paul Churchland are famed eliminative materialists of this kind.[9] Materialism or physicalism are related philosophical stances that consider matter, or the "physical," to be all that exists. A lot hinges here on what we mean by *matter* and *physical*, and there is a terrain of philosophically nuanced positions in this area. At the more naive end of the spectrum, we might have materialists who believe matter, as in the particles studied by physics, to be a solid substance and the only thing that truly exists. A more sophisticated physicalist might accurately claim instead that we never actually come into contact with some substance called matter; physicists only quantitatively describe the behaviors of certain aspects of the world. The physicalist may believe that seemingly nonphysical processes, like mind, are dependent on these processes that physicists study but without committing to a stance on exactly what physical processes are. Emergent phenomena like biology, minds, and societies can exist, but they are all built on top of the processes of physics. This is known as nonreductive physicalism.

Eliminative materialists, though, believe that the only thing that exists is matter and that the mental states we believe we possess do not actually exist. Take the example of a belief. We typically feel our beliefs to have meaning or content; that is, they usually refer to something. We also typically ascribe some causal power to those beliefs. If I believe there is an apple pie in front of me because I can smell it, and I believe I am hungry because I can feel it, then our commonsense explanation of what is going on when I approach and eat some of the pie might involve claims that I held beliefs about the existence of the pie and my hunger, and those beliefs resulted in the action of approaching and eating it. To the eliminative materialist, this talk of beliefs is just some folk-psychological idea of what is going on that doesn't relate to what is really going on, like someone believing an evil spirit is the cause of someone's mental illness. In reality, all that happens is odorant molecules enter the nose and trigger complicated physical processes in the brain that result in the movement of muscles.

A related idea is the proposal that conscious experience itself is an illusion. According to the illusionist philosophers Daniel Dennett and Keith Frankish, when we introspect and feel that there is a certain quality associated with an experience, such as the unpleasant feeling of pain resulting from an injury, no such quality actually exists.[10] Rather, the brain pulls off a trick where it gives the impression that the purely physical signaling of damage to the body has a quality that it doesn't actually have. You might ask how it could give such an impression if there was no space of experience in which the impression could arise. For many, this is a nonsensical position.

There is another path for those who are convinced that science and philosophy cannot connect consciousness to matter in principle. It is known as mysterianism and is associated with philosopher Colin McGinn.[11] A mysterian believes that there are some things that we simply can't understand and that consciousness is one of them. We are humble primates, and there are limits to our knowledge. We wouldn't expect gorillas to understand the stock market, so why do we insist that we must be able to understand all aspects of reality? For many, however, to assume that we are incapable of understanding consciousness because we haven't done so yet is to give up the fight prematurely.

If we find ourselves in the position that none of these proposals seem to satisfactorily explain the existence of consciousness yet we are not ready to throw in the towel and assume the answer is simply beyond us, then we have to consider the possibility that there is something wrong with how we are thinking about consciousness and the world in which it exists. If we have taken a wrong turn, then it might help to retrace our steps to see where a mistaken assumption or two might have crept into our thinking.

Our current dominant paradigm is known as reductionism. According to reductionism, all that truly exists is the world of physics; only the microphysical level of subatomic particles and atoms is considered to contribute to our picture of how nature works. According to this story, everything that ever happens is entirely due to the activity of these particles. Where we see things happening that seem to not be microphysical particles, such as chemical reactions, the physiology of life, the activity of the mind, or societies and civilizations, they are not really part of the story of reality, despite their appearances. They are epiphenomenal, just a by-product of the unfolding dynamics that physics studies. Your goals and plans do not actually contribute to how your body moves through space. Instead, there is only a vast mechanical process consisting of particles interacting—the idea that you have ever made anything happen in your life is a mere illusion.

According to this view, if we could track the location of every particle in your body, then we wouldn't need neuroscience to explain your behaviors. Brains do not actually achieve anything themselves; they are just the large-scale appearance of microphysical processes, and it is at this microphysical level that all the action is happening. This idea was first proposed in 1814 by Pierre-Simon De Laplace, who argued that a vast intellect or demon with knowledge of the positions and motions of all the particles in existence would be able to precisely predict the unfolding of the universe based on this information alone.[12] This idea that everything can ultimately be explained by the deterministic motion of particles moving through space can be traced back even further to Newton.

Newton's equations of motion do an incredible job of describing how one or two objects move through space. The problem is that these elegant laws break down once you add a third participant to the interaction, a fact known as the three-body problem. Given that our current scientific picture of the world is based on the idea that everything that ever happens is due to laws such as these, you would be right to be shocked by the fact that they only apply to such a narrow range of phenomena in our universe. Try applying Newton's laws to the activity of any molecule in your body, and you will find that they fail miserably. The idea that the reductionist paradigm explains our universe relies on putting on blinders that exclude most of what we care about, including all of mental life and the whole of the living world. We must update our thinking to be able to account for these phenomena in a scientific fashion.

THE CONTEMPLATIVE'S PERSPECTIVE: BEGINNING WITH CONSCIOUSNESS

Become conscious of being conscious. Say or think "I am," and add nothing to it. Be aware of the stillness that follows the "I am." Sense your presence, the naked unveiled, unclothed beingness. It is untouched by young or old, rich or poor, good or bad, or any other attributes. It is the spacious womb of all creation, all form.—RAMANA MAHARSHI[13]

What would be the problem if it turns out that consciousness and the physical world are entirely distinct? What if it is just not possible to account for consciousness in terms of physical processes? This stance is known as dualism because it posits the existence of two fundamentally different aspects of reality: the material and the mental. Perhaps the figure most associated with dualism is Descartes.[14] He proposed a picture of reality in which nature is entirely unconscious and mechanistic, while human consciousness is attributed to a divine, immaterial soul. Descartes thought that mind and matter were truly separate but that the divine mind could influence the material body via the pineal gland, a small structure in the middle of the brain. Dualist accounts of this kind face a major problem, however. If these two substances are entirely separate, then how could they possibly interact with each other? If they interact in lawful ways, then that means they can be considered part of a single greater system and are therefore not fundamentally separate. Due to this paradox, dualism has largely fallen out of favor since Descartes's time.

In trying to understand how matter and mind relate to each other, we are looking for a monistic view of reality, not a dualistic one. Physicalism is a kind of monism, where only one thing—the physical—exists. Others have gone all the way in the other direction, proposing that only mind truly exists, with matter being a secondary aspect or even an illusion. There is certainly one sense in which the mental occurs prior to the physical.

Before you ever learned a single scientific concept, there was an experience of existing. You might have looked out at a world of experience in which a teacher or a textbook informed you that reality was made of things called particles, but you had never seen such a particle. How could this thing that you have never experienced, a mere idea from your perspective, be more real than your direct experience in this moment? From your perspective, experience is more fundamental than matter.

It is certainly true that consciousness is more real and fundamental in this personal sense, in terms of the situation that we find ourselves in when we want to understand the world and our place within it. In fact, this experience is the only thing you know for certain. In this moment, you experience what appears to be a world around you, but if we were in a pedantic or skeptical mood, we might wonder if we could be wrong about that. You might be dreaming right now, or you might be plugged into a simulation, as in the film *The Matrix*. The one thing you can be certain of is consciousness. It is the one thing you cannot doubt, as even the act of doubting requires its existence. This is what Descartes was referring to when he wrote the famous sentence "*Cogito, ergo sum*"; "I think therefore I am." The existence of consciousness is the one thing we cannot doubt.

When we begin thinking about how experience relates to the rest of the natural world, consciousness comes first. We are first conscious, then we come together to figure out how the world around us operates, with the ultimate hope of accounting for the fact of experience that we began with. There is another sense in which consciousness can be thought of as coming first when thinking about this issue. The oldest known spiritual texts in the world, the Vedas and Upanishads that form the basis of Hinduism, deal with the relationship between consciousness and the world. The people who orally transmitted and later wrote down these texts lived in the Indus River Valley, from which we get the words *India* and *Hindu*, between 1500 BC to 600 BC. It is written in these texts that, through visionary experience, one can discover oneself to be a manifestation of Brahman, the totality of the universe, like a drop of water that is not truly separate from the rest of the ocean. If you are ultimately the same thing as the rest of the universe and what you feel yourself to be consists of conscious experience, then perhaps the true nature of the universe is consciousness. After all, as a part of the universe, shouldn't you be able to interrogate yourself and discover what its nature is? If this were the case, we could conceive of all existence as like a dream in a cosmic mind. Psychology would be the most fundamental science, not physics.

This idea is most associated in the West with George Berkeley. This eighteenth-century Irish philosopher was a Christian bishop who denied that anything actually exists in the world beyond our conscious experience of it.[15] When you see a chair, there is the mental image of the chair, but there isn't actually a physical thing out there in the world. According to Berkeley, the whole of existence is made of mental images, of ideas. It is from the word *idea* that this position gets its name, idealism. From this perspective, consciousness not only comes first in our personal experience of the world, but it is understood also to be at the origin of the universe because it is the very nature of existence. This perspective attempts to account for where consciousness fits into our picture of reality by claiming that it *is* reality. What we call matter is merely how this cosmic consciousness appears when we look at it. We can measure it with quantities, as we do with science, but those are mere measurements; the thing we are measuring is an aspect of a single universal consciousness.

On the surface, this perspective may seem more spiritual or religious in nature rather than scientific, especially with its association with ancient Hinduism in the East and a Christian bishop in the West. Despite this, a very similar idea is now experiencing a resurgence in scientific circles. The seeming insolubility of the hard problem drove its formulator, David Chalmers, to a position known as panpsychism, in which consciousness is seen as a fundamental aspect of our reality. Rather than attempting to understand how consciousness came to exist, this approach posits that consciousness has always existed wherever there is matter; they are two sides of the same coin. The subtle difference to idealism here is that, rather than matter being a secondary appearance in a more primary cosmic consciousness, matter and mind are seen as both existing in an equally fundamental manner.

While it may seem radical, this position has a number of highly respected advocates. At the beginning of the twentieth century, this stance was proposed by one of the most important philosophers of the last one hundred years, Bertrand Russell.[16] More recently, it has been championed by philosopher Galen Strawson, alongside fellow philosophers Chalmers and Philip Goff.[17] Even Francis Crick's collaborator in the investigation of the brain basis of consciousness, Christof Koch, has embraced a scientific form of panpsychism called integrated information theory, developed by Giulio Tononi, an approach that commits to a worldview in which even atoms are conscious.[18] Cognitive scientist Donald Hoffman has even argued that underneath physics, at the most fundamental level, reality is made of interacting "conscious agents."[19] I save the exploration of these ideas for a later chapter. For now, let's just say that these positions come with their own problems that make them insufficient solutions to the problem at hand.

SOLVING THE HARD PROBLEM

The head spins in theoretical disarray; no explanatory model suggests itself; bizarre ontologies loom. There is a feeling of intense confusion, but no clear idea about where the confusion lies.—COLIN McGINN, ON THE MIND-MATTER GAP[20]

Science and philosophy both attempt to provide a unified account of how the world operates and how its different aspects relate to each other. As philosopher Wilfrid Sellars put it, "The aim of philosophy, abstractly formulated, is to understand how things in the broadest possible sense of the term hang together in the broadest possible sense of the term."[21] This applies to science, too, which began its life as a branch of philosophy concerned with the natural world, known as natural philosophy. *Nature* here refers to all physical existence and can be contrasted with the supernatural, rather than referring to nature in the ecological sense.

In this picture, the idea of a fundamental division in reality poses a problem: We can't come up with a complete picture of our world if an unbridgeable fissure exists between matter and consciousness. As a result, most people who think about consciousness subscribe to monism rather than dualism, the idea that reality is really one thing rather than being split into two separate substances. We must agree on a particular monistic picture of the world if we are to understand what we are in our essence, both as conscious things and as living things.

The gap that exists in our understanding between matter and mind is not a monolith; it consists of multiple submysteries. The first, and perhaps most important, is how the qualitative can exist in our seemingly quantitative world. A second issue is how such qualitative experiences can have meaning, both in terms of how they can hold significance for us and in how they can be about something other than themselves. Finally, we must explain how experience affects the world, what causal power it has, as only by making some change in the world can it be said to truly exist from a scientific perspective. We will come across other problems on the way, such as the nature of the self.

To understand consciousness, we must start in a place that will be surprising to some and obvious to others: with the experiences of contemplatives who have become experts in examining consciousness. Today, scientists are often wary of the claims of such people, as they typically rely on subjective reports and are therefore largely unverifiable. While this is a concern, other researchers have taken on the task of attempting to verify these claims using science. By looking at consciousness using the synthesis of our best objective and subjective approaches to the mind, we will put ourselves on an optimal footing for the journey ahead.

TWO

DEMYSTIFYING MYSTICISM

Beyond Separation

Standing on the bare ground,—my head bathed by the blithe air, and uplifted into infinite spaces,—all mean egotism vanishes. I become a transparent eye-ball; I am nothing; I see all; the currents of the Universal Being circulate through me; I am part or particle of God.—RALPH WALDO EMERSON[1]

NATURALISTIC MYSTICAL EXPERIENCE

On an ordinary spring day in 2003, I was riding the number 74 bus up East Hill in the English town of Colchester, next to the town's Norman castle. I was thirteen years old, and as a result of the Catholic culture in which I'd grown up, I was frantically ruminating about the concept of blind faith. All the authority figures in my life were telling me that I lived in a universe in which I was morally obliged to believe in a supernatural creator in spite of a complete lack of evidence. The kicker was that my inability to do this would result in unimaginable torture in hell for all eternity.

This was the single most important issue in all of existence, as far as I was concerned, and it was therefore imperative that I find a way to believe in something that I had not even the faintest reason to believe in. Why had this omnipotent creator given me an empirical mindset that made it impossible for me to believe in its existence? Why would it torture me forever for a failure that it itself had built into the very bedrock of my mind? I spiraled further and further into this train of thought, becoming increasingly overwhelmed by the cosmic consequences of my inability to solve this impossible riddle. Then, as this existential anguish approached the point of being unbearable, my thoughts stopped. Completely.

I looked out of the window of the bus at the world that I had been totally unaware of until that moment. The wind rustled the leaves of the trees around the castle. Behind both, the sky shone a brilliant blue. With thoughts eliminated, the everyday world was now an extraordinary vision of natural perfection. The entire scene seemed to be effortlessly offered up by existence itself, requiring no act of observation from a separate observer to be perceived. Everything in experience appeared to be self-illuminating. Nothing intervened in the seeing; it was perfect intimacy with the seen. All appeared as simultaneously transparent and unimaginably light yet also full, potent, and maximally profound. My mind no longer imagined "James" to be something apart from the rest of existence. With the cessation of this imaginative act, I was fully absent, and the world was fully present. It wasn't that I felt one with existence—rather, it had suddenly become profoundly, earth-shatteringly obvious that there only *was* existence. With no thoughts to judge whether the appearances were good or bad, better or worse, right or wrong, they were revealed to be absolute perfection. Everything was exactly where it needed to be.

The serenity in that moment was beyond anything that can be put into words. All suffering had evaporated. What was given in the present moment was so radically absorbing that beholding it was satisfying in a way that I had never known. It was like an infinite disclosure of the ground of existence, an eternal, unending revelation of that which is beyond intellectual understanding. Ideas of past and future were seen as imaginative acts, occurring in the inescapable, totalizing present moment. With the distortions of thought eliminated, the indescribable obviousness of *what is*, here and now, was fully on display. Rather than feeling like an altered state, it felt like quite the opposite. It felt as if I had spent my life until that moment in an unreal state in which I had identified with the abstract conceptual contents of the mind, a state of struggle and disconnection that had finally been dispelled, and what remained was utter clarity. This shift in consciousness is often called "awakening" in spiritual circles, and it did indeed feel like waking up from a dream, which is how it still feels to this day.

What happened on that day? We know that the complex mental world of meaning that we humans create isn't fundamental. Physicists aren't running around looking for the basic laws of the universe that explain why you prefer chocolate to vanilla ice cream. What I discovered on that day is that it is possible to experience in a way that is beyond our usual mental constructs. What's more, the suspension of these mental processes is associated with a suspension in suffering. Such claims often evoke feelings of skepticism, and understandably so. As a scientifically minded individual, I personally would have had no appetite for the intellectual terrain of spirituality if my mind hadn't collapsed into a mode that revealed these kinds of states to exist. Given how much I had suffered because of religious dogma, these outlandish-sounding claims are the last thing I would have chosen to advocate for. Having recognized the reality of these states, however, I became dedicated to exploring them in a way that is maximally scientifically and philosophically grounded.

Scientists are now studying these states of consciousness, terming them *mystical experiences*. This term is not intended to indicate that they are somehow inherently ethereal or magical seeming but that they are associated with the religious tradition of mysticism, in which insight is derived experientially rather than through other modes, such as the studying of scripture. I cannot stress enough that there is nothing inherently supernatural about the mystical state; in my case, it felt profoundly naturalistic and compatible with science.

The way of seeing the world that dawned on me that day has deeply influenced how I understand both the nature of consciousness and its place in reality. The core insights that are relevant here relate to the experience of feeling separate objects to exist, the feeling of being a conscious self, and the belief that the world is fundamentally made of a solid material substance. It now feels obvious to me that separation, self, and substance are all mental constructs without their own independent existence. This is a radical departure from our commonsense intuitions about what we are and the way the world is, so I begin by exploring the ways in which these claims fit with what we know from science. Talking about insight derived from exotic states of consciousness will surely alienate some who find this departure from our everyday intuitions too much to swallow. The insights that arise from these states, when combined with scientific skepticism and philosophical rigor, offer a remedy to our current intellectual malaise and so must be tackled head on.

ONE UNIVERSE

*A human being is a part of the whole called by us universe, a part lim-
ited in time and space. He experiences himself, his thoughts and feeling
as something separated from the rest, a kind of optical delusion of his
consciousness.* —ALBERT EINSTEIN[2]

The physical world is fundamentally undivided. The energy within life-
forms is channeled from organism to organism in relentless food chains.
Continental plates descend into the earth's core to be melted down into
magma, which will reform as fresh land. Even stars die, eventually mountains crumble, and planets are destroyed, all being recycled in the circular
channels of nature. It's not only the wider physical world that is a messy
mixture of this kind; so is our immediate environment. We exist in an ocean
of intermixed vibrations traveling through the air, our bodies drenched in
solar radiation that bounces off countless surfaces around us. Even we are
fundamentally open energetic processes that depend on the consumption
of energy sources in our environment to keep the wheels of metabolism
turning. Before our minds begin to make sense of it, the world is a single
riot of energy.

The physical world may be continuous, but it typically doesn't feel that
way to us. It is true that, from your perspective, some physical forms can
be engaged with in a way that others can't, leading to you perceiving then
as separate objects. For you, a rock that you can hold in your hand is something that you could feasibly use, and so it is categorized as a distinct thing.
When it comes to the rings of Saturn, however, it's a different story. Is
this whirling collection of matter one thing? Should you instead say there
are many rings? If so, how many? Given its lack of utility to you on your
scale, you may simply remain agnostic, which is wise, as there is no truth
of the matter of the actual number of objects present. Objects and their
boundaries are things that we perceive from our perspective, not things that
objectively exist.

Is it not true that a single solid rock is in some sense more of an object
than the rings of Saturn? In the objective description of the universe, even
rocks do not exist as truly separate objects. Years ago, the rock was a continuous part of a mountain, and in the future, it will become sand. Without

us to perceive it as a rock at a given moment, there is only a continuous physical process. From the perspective of physics, the rock is merely a collection of atoms that will ultimately disassemble and mix with the rest of existence. It is not even distinct from the air around it. Rocks are continually weathered by the wind, and when a limestone rock is weathered by the rain, parts of it are released into the air as carbon dioxide gas. Without us mentally drawing borders between what we call the rock, the air, and the rain, no such distinctions exist. The physical world is continuous; it is only in our minds that it is felt to consist of separate objects. In his book *The One*, particle physicist Heinrich Päs makes the case for this unitive vision of physical reality, showing how this understanding becomes even more relevant when we move to the subatomic scale.[3]

What about living things? Are you and I not separate things? As I explore throughout this book, while living things do function as distinct agents, we are processes rather than fixed objects. We constantly exchange gasses with the world around us when we breathe, and we build our bodies from the food we consume from outside ourselves. While we perpetually distinguish ourselves from our physical environment for the time that we are alive, we never fully succeed in separating from the rest of existence. This is fortunate, as we wouldn't be able to survive if we did. We and all living things are like whirlpools in reality—we are not separate things.

In ancient Greece, confusion around the nature of objecthood of this kind was explored in the ship of Theseus. We are asked to imagine that over time every single part of the ship wears out and is replaced. Eventually no part of the original ship remains. Can we say that it is still the same ship? We can understand what is going on here by appreciating objecthood to be a mental construct, a perceptual overlay that we superimpose on a continuous physical reality in which no truly separate objects actually exist.

THE FICTION OF OBJECTS

Nature is a self-regulating system, in which everything is interconnected, interdependent and in a constant state of flux.—FRITJOF CAPRA[4]

To navigate a messy, continuous world, the brain divides the huge mass of sensory data that streams into it into distinct categories that are useful for survival. This is the process that is responsible for the perception of object-hood. In order to survive, we must detect patterns in the sensory signals we receive that could be relevant to us and discriminate between different patterns. Imagine walking in the woods and suddenly seeing a long, thin shape against a background of leaves on the forest floor. Your brain picks out that its shape is not physically continuous with the leaves and perceives it as a distinct form that can be interacted with in that moment. It could be a piece of wood or a snake, and your brain needs to immediately determine which. The visual input it receives needs to be assigned to the category *snake* or *not snake*, even if the signals your eyes are receiving are ambiguous. Resting at a 50 percent possibility it could be a snake will not serve you well for your survival. Instead of accurately representing the sensory signals it receives, the brain assigns categorical interpretations to these signals and projects this divided structure back out onto the world.

In some cases, it is obvious to us that our perception is structured by categories. A gardener who is constantly fighting a plant they do not want in their garden may assign this innocent piece of vegetation to the category of weed. The plant is not objectively a weed and may be seen as desirable by others, but the gardener will likely have their perception colored by their attribution of this category to this little green organism. This in fact happened to me when I moved onto the homestead in the mountains of southern Portugal, where I live. My first spring there, hundreds of beautiful cistus flowers blossomed over the land, pure white circles of delicate crepe-paper petals that looked like punctures in reality. I commented on their beauty, and a local told me they were a weed that I would have to keep under control. In real time I felt my vision of the flower darken, and rather than being absorbed in the beauty of its petals, I suddenly was noticing how annoyingly sticky the sap on its stem was when touched.

During development, your brain learned how to chunk reality into the useful concepts that you use for perception. For example, we do not perceive speech sounds as they physically are but sort them into categorical bins. If you use a computer to slowly morph between the sounds *ba* and *da*, you will not experience a smooth transition between the two, hearing a perfect mixture of the two in the middle. Instead, at some point on the continuum, your perception will shift from hearing *ba* to *da*.[5] This is unlike what happens when we mix such colors as blue and green, where we readily are able to perceive a bluish green when they are equally mixed. This categorical perception of sounds would be like perceiving vivid blue with no green at all, even when it is a 60/40 mixture and your perception suddenly flipping to a vivid green with no blue when the ratios are reversed.

Why does this happen with speech sounds? When speaking English, the sounds *ba* and *da* are produced more often than a mixture of the two. Given that our mouths are fallible and our speech is imperfect, it makes sense for the brain to perceive what we believe the speaker was intending to say based on previous experience with the English language, rather than hearing the sloppy sounds exactly as they are. Through development, the neural networks of your brain come to capture the landscape of the likely and unlikely sounds of your language, exerting a gravitational effect on the sounds you hear, to pull them into the most likely categories. In this way, a messy world is cleaned up into an experience of distinct objects that can be usefully engaged with.

This understanding can help us make sense of why we perceive objects when none actually exist. There are patterns in reality that are more or less relevant to our survival, and our experience becomes shaped to capture the patterns that are of importance to us. What we must appreciate, however, is that this process goes above and beyond in giving the impression that these patterns reflect genuinely separate objects. This would be like writing your name in clay and then thinking that you can pull out the letters from the substance you imprinted them on. There is no way to separate any object from the rest of the universe, even for a moment.

HALLUCINATING REALITY

You look up at random patterns in the clouds and perceive the shapes of animals. Someone on the other side of the world sees the face of Jesus in their burnt toast. The internet erupts in a dispute about the color of a dress. These moments illustrate a truth about our everyday experience: Our consciousness is not a transparent window on reality; what we perceive is constructed by our brain.

Our experience doesn't typically feel constructed. It usually feels like we effortlessly look out on an outside environment consisting of many different objects with us at the center. This is as it should be: You evolved to perceive a world around you so that you could function in it, not so that you could reflect on the nature of your own mind. We so fully take for granted that we are directly in touch with reality that we are shocked when we are confronted by the fabricated nature of our experience of the world. In 2018, a sound file went viral of a voice saying either *yanny* or *laurel*. This difference in perception did not arise from the sound waves themselves but from the minds of the listeners, revealing the way in which we shape what we perceive without realizing it.

For many years, the process of perception was conceived of by scientists as a process of extracting information from the world around us. If this view is correct, then we should see the world as it is. Let's see how such a scenario would work in the brain, taking vision as an example. This model of how perception could work begins with visual information coming into the brain via the retina. Once in the brain, neurons detect whether simple visual features are present, such as edges of different orientations. The output of these neurons is then passed on to another set of neurons that detect whether more complex features are present, such as curves. This process proceeds up a hierarchy until we arrive at neurons that detect highly specific images, such as your grandmother's face. This specific analogy gave rise to these hypothetical high-level feature detectors being called "grandmother cells."[6]

From this perspective we would assume that most of the connections in the brain send signals away from the sensory receptors, such as those in the retina, and toward the next level in the hierarchy. Interestingly, this is not the case: The majority of connections in the brain actually flow in the opposite direction. An answer to the mystery of why this is has arisen in the last few decades, with what is now known as the predictive-processing perspective.[7] Rather than perception proceeding through a process of extracting features that exist out there in the world, ready to be detected by the brain, the predictive-processing account suggests that the brain actually first generates its best guess about what is out there in the world, predicts which signals it would receive if this guess were correct, then checks with the actual incoming evidence in order to update its guess. It turns out to be vastly more energy efficient and effective for the brain to infer the nature of the world around it through clever guesswork rather than laboriously extracting a picture of how the world really is from incoming sensory signals.

According to the predictive-processing approach, our experience of the world around us is a "controlled hallucination." It is a hallucination in the sense that it is internally generated, and it is controlled in the sense that its contents are kept in check by the information that the brain receives via the senses. We do not see the world directly; we instead experience an internally generated simulation of a world. What's more, the function of this simulation isn't even to accurately capture what is going on in reality. Instead, its purpose is to help us navigate effectively. If you see a threatening person in the woods at night when in reality all you are looking at is a tree, then your brain has done a good job of displaying what is most appropriate for your survival. Better to be on the safe side and detect a threat where there is none than be complacent and run the risk of dismissing an attacker as a harmless tree. The point is that we do not see the world as it is. What we take to be the world around us is in fact a useful map for navigating existence.

If it were the case that we perceived the world directly as it is, then there wouldn't be much we could do to alter our experience of the world. Given that perception is an inherently creative act, however, it is possible to profoundly alter the contents of one's hallucination of reality.

BEYOND DIVISIONS

I entered the meditation hall for the tenth day in a row and sat down. The meditation bell rang, and I closed my eyes. My body was nowhere to be found; there was only a pointillistic tempest of energy, with no experience of boundaries anywhere. Nothing had gone wrong. Such states are part of the journey when engaging with deconstructive meditation techniques. In such approaches, attention is used to inspect experience with greater and greater precision, leading to multiple insights. One of these insights is impermanence, that everything in experience is always in a state of flux. The flip side of this is that any sense of solidity is only a temporary mental construct that will ultimately dissolve into the currents of change. Another is that, no matter how hard you look, you will never find a fixed boundary between any apparent objects in experience.

It is not necessary to spend days in silence on a meditation retreat to have this insight, however. The experience of separation between objects turns out to be a very thin overlay that can be seen through in an instant. Imagine looking at the *Mona Lisa* in a gallery. You clearly see a woman with a landscape in the background. Next you turn to see Van Gogh's *Starry Night*. This would technically involve you teleporting from Paris to New York, but ignore this point for now. Thanks to Van Gogh's impressionistic brushwork, you notice that, despite being able to pick out cypress trees, hills, and buildings, you are really looking at paint on a canvas, with these seemingly separate objects only implied. When you go back to look at the *Mona Lisa*, you now see a continuum of paint and realize that the image of the woman doesn't objectively exist on the canvas. In reality there are only splotches of paint, the woman only exists in your mind. It would have been possible to have this insight immediately without needing to deconstruct the objects into paint forms, as the Van Gogh invites you to do, but training oneself to make the distinction between apparent objects and the medium they appear in can be a useful intermediate step.

This process is analogous to the practice of deconstructive meditation, while the immediate recognition is analogous to my mystical experience on the bus. Much as the apparent objects in these artworks are formed in the medium of oil paint, the apparent objects in our experience are formed in the medium of consciousness. It can be challenging to recognize this fact in the midst of day-to-day life, but in this type of meditation, it is possible to focus your attention precisely enough to inspect whether the boundaries that we think exist between perceived objects are really there. When we do this, we find that there is only continuity in experience, like Van Gogh's fluid brushstrokes that only ever give the impression of separate objects.

Another way to see through the conceptual division between objects is through the use of the class of chemical compounds known as psychedelics. Classical psychedelics, such as LSD and psilocybin, the active compound in magic mushrooms, act on the brain to disrupt the ability to enforce conceptual divisions between perceived objects. As a result, your experience of reality can begin dissolving into an impressionist Monet painting—or into Munch's *The Scream* if you are reckless and ill prepared. Either way, the melding of the contents of experience can lead to the dramatic realization that what we take to be the world is actually occurring in consciousness. It is a simulation of the world rather than the world itself.

THE NATURE OF INSIGHT

Our normal waking consciousness . . . is but one special type of consciousness, whilst all about it, parted from it by the flimsiest of screens, there lie potential forms of consciousness entirely different. We may go through life without suspecting their existence; but apply the requisite stimulus and at a touch they are all there in all their completeness. . . . No account of the universe in its totality can be final which leaves these other forms of consciousness quite disregarded.—WILLIAM JAMES[8]

Mystical experiences are states in which conceptual divisions are seen through and existence is perceived to be whole and undivided. In what way is the shift in perspective that can occur with a mystical experience relevant to our scientific understanding of consciousness? Science is conducted by fallible humans, humans who hold their own biases about the nature of the world around them, biases that filter into consciousness science itself. Typically, we experience ourselves as separate from the rest of the world, and we experience that world as consisting of solid, material objects. As a result of our experience, many of us hold beliefs in the existence of a conscious self, separate objects, and solid substance. The mystical experience can dramatically bring these assumptions into question, seeming to reveal all three to be mental constructs with no objective reality.

The idea here is that nonordinary states of consciousness of this kind may be an antidote to some of the unconscious assumptions we bring to the table when thinking about experience. For this to be the case, the perspective on reality that comes online with such experiences would need to be seen as a source of genuine insight, rather than as a source of hallucinatory delusions. Those who engage with experience in this way do typically feel the rigorous exploration of consciousness to be a pursuit that can lead to knowledge and wisdom. Buddhist monks do not dedicate their lives to countless hours of meditation simply to pursue delusional, trippy experiences; they believe they are doing so in the service of clear seeing, insight, and experiential liberation. What's more, we know from the field of psychology that our minds do not operate like ruthless logical truth seers but are riddled with evolved biases and delusions. The idea that we might be able to explore the nature of these delusions and see beyond them should not seem so outlandish.

A major insight that occurs time and time again in the mystical state is that existence is inherently undivided, or whole. How could it be possible to directly perceive features of reality through such insight experiences? One way to understand this possibility is to appreciate that, in these cases, any new knowledge typically arises from seeing the false nature of existing beliefs. It's not so much that information about the unity of existence is mysteriously transmitted to a person during such an experience. Rather, our ability to imagine that the world is divided into separate things breaks down. The resulting insight is not so much that the world is unified but that it is not actually divided as we had thought. This may seem a subtle difference, but it is important in showing that nothing spooky is going on here. The insight does not arrive from somewhere outside the individual but occurs as the result of purely internal processes in the brain.

Even if this is the case, how is it possible to gain understanding through direct experience? We may be very familiar with rational and logical approaches to learning, as when we study a topic by reading a book or are taught by another, but is it also possible that something can be learned in a way that is nonrational? The answer is an unequivocal yes; there are multiple other ways to acquire knowledge beyond the language-based terrain of the logical and rational. These nonrational modes of knowledge acquisition include things like learning to ride a bike or learning to recognize someone's face or any object, for that matter. You don't say, "Given the size, shape, location, and other properties of that object, I logically conclude that it must be an apple." You wordlessly, intuitively apprehend what it is with no mental effort at all. The way we come to know the world through perception is nonrational; it is intuitive.

The vast majority of processing in the brain is of this nonrational kind. Most animals manage to acquire an understanding of the world around them in a nonrational way that doesn't depend on linguistic analysis. The same is happening right now to give you the experience of the world you inhabit and of yourself within it. A vast amount of information flows through the networks of the brain, being sculpted to arrive at the most appropriate conclusion of what is going on through an intuitive mode of analysis. There is no mystery of how this could be possible, either. Modern artificial intelligence (AI) systems were directly inspired by the brain's ability to build associations based on experience. The processing performed by such a system is not based in the rational analysis of logical propositions. These AIs function primarily as intuition engines, picking out patterns in vast arrays of data. Rationality rests on a huge foundation of nonrational knowledge acquired through perception. Intuitive knowledge is so fundamental to our experience of the world that we barely notice it for what it is.

When one intuitively apprehends the undivided nature of the world in the mystical state, nothing supernatural is occurring. It is just like seeing an apple; it is simply recognized. Of course, just because you think you see an apple in front of you doesn't mean there really is one. You might be dreaming or hallucinating. Rigorously testing mystical insights scientifically and philosophically is therefore crucial, as their validity cannot be simply taken for granted. I show in the next chapter that research into the constructed nature of the self is bearing out these claims.

FROM MYSTICISM TO MONISM

All the dogmatists have been terrified by the lion's roar of emptiness. Wherever they may reside, emptiness lies in wait!—NAGARJUNA[9]

What does an apparent object look like when it is no longer perceived as a thing, separate from everything else? Consider a flower. We apply the concept "flower" to certain plants, and our mind projects the impression of "essence of flower" onto that particular natural process. But consider a full description of what the flower is and why it is that way. Such an explanation would require us to invoke the existence of bees to account for the colors of the petals and the existence of gravity to explain the stem's vertical growth pattern. The fractal patterns of its roots imply the existence of rainclouds. Its broad green leaves contain within them the existence of the sun. At the end of the analysis, there is no essence of flower to be found apart from all these processes that implicate the rest of the universe. In the words of Vietnamese Thiền Buddhist monk and poet Thích Nhất Hạnh, "A flower, like everything else, is made entirely of non-flower elements. The whole cosmos has come together in order to help the flower manifest herself. The flower is full of everything except one thing: a separate self, a separate identity."[10] In Buddhism, this is known as emptiness and is associated with the influential Indian philosopher Nagarjuna and his work *Mūlamadhyamakakārikā, The Root Verses on the Middle Way.*

Rainbows are an excellent illustration of emptiness and interdependence. While rainbows do exist, there is no physical thing present where we perceive the rainbow to be located. There is no essence of rainbow—it is empty. What there is, however, is an interdependent or relational process, one that involves humidity, sunlight, and an observer in the correct location. It turns out that everything in existence is like a rainbow, relational and empty of an intrinsic, independent essence.

When we stop conceptualizing, things are perceived to be empty of a separate existence, they are instead perceived to exist through the influence of everything else in reality. This is known as interdependence, interbeing, or relationality. Think of the way in which an ecology exists. An ecology is a networked structure. Without the connections between the elements that make it up, the ecology would not exist. An ecology is not defined by the creatures that exist within it but by the interactions between them. We might think that the organisms are truly separate individuals, but we can apply the same analysis there. Many creatures, including us, are symbiotes. The mitochondria that exist in every cell in our bodies are an organism from a different kingdom that our ancient multicellular ancestors teamed up with. If we drop our usual way of seeing, we can see ourselves not as individuals but as intersections of a vast array of natural processes. From this perspective an ecosystem like the Amazon rainforest can be beheld as a seething riot of energy relating to itself, not a collection of distinct organisms.

Emptiness and interdependence are not just features of physical things but also of the contents of experience. Imagine exiting a dark cave into the bright midday sun. It may feel as if we perceive darkness and brightness as possessing their own independent, separate qualities, but in reality, they mutually depend on each other. Bright is the same thing as not dark, and vice versa. Your experience of brightness and darkness, upward motion and downward motion, and high-pitch and low-pitch sounds all exist in a way that is dependent on their opposite. No single experience can stand alone without relating to all the other possible experiences that you could have. Philosophers have speculated that we could conceive of a single quality, termed a *quale*, such as the color red. If all you could experience was red, however, the lack of variation and structure would be identical to having no experience at all. The apparent individual qualities of conscious experience are as empty of essence as the objects we perceive around us.

When the categorical mental constructs that give us the impression of living in a world of separate objects are dismantled, we are left with a vision of reality as inherently whole because of its interwoven structure. This way of seeing the world fits perfectly with the philosophical stance of monism, the commitment to reality not being fundamentally divided into different substances, and the perspective we must take to reconcile matter and mind. This commitment to monism and the understanding that no apparent thing is truly separate from the rest of reality are both essential insights for understanding consciousness and its place in existence. If there are no true separations in reality, though, then what are we to make of the most fundamental seeming separation of all, the one that appears to lie at the core of consciousness? I speak here of the seeming separation between what appears to be an experiencing subject (i.e., you) and the world.

THREE

THE ILLUSORY SELF

Beyond the Subject

We sit together, the mountain and me, until only the mountain remains.
—LI PO[1]

I walked through the olive grove as the midday sun pinned the neat shadows of the trees down to the ground beneath them. Birds sang in the branches. The wind rustled the leaves. The gravel path crunched under my feet. In that moment, something was noticed that was simultaneously totally ordinary and yet extraordinary at the same time—experientially, I was not there. The body was still there, but it was completely part of this organic tableau of life-forms and minerals, from rocks to birds to plants to primate. Only the totality of the olive grove remained. No matter where I looked, there was nothing separate from the scene. There was no need to separate the trees or the insects or the birds, and if they could be part of an olive grove, then why not a *Homo sapiens*? The body walked on, the sound of the gravel underfoot arising in pure, simple clarity, appearing for no one.

WHAT IS THE SELF?

For my part, when I enter most intimately into what I call myself, I always stumble on some particular perception or other, of heat or cold, light or shade, love or hatred, pain or pleasure. I never can catch myself at any time without a perception, and never can observe any thing but the perception. . . .

I may venture to affirm of the rest of mankind, that they are nothing but a bundle or collection of different perceptions, which succeed each other with an inconceivable rapidity, and are in a perpetual flux and movement. —DAVID HUME[2]

What is conscious of these words in this moment? A common answer would be "I am." This *I* is the self, the subject that we feel *has* experiences. The intuition that we possess a self that is capable of having experiences is very widespread, even in philosophy and science. Many scientific theories of consciousness are centered around the idea that we possess a self-like mechanism in the brain, a mechanism that is capable of bestowing the light of consciousness on the signals that exist in other parts of the brain. It is therefore essential that we investigate whether this conscious self is what it seems to be.

In his classic work *Consciousness Explained*, philosopher Daniel Dennett describes this self-based idea of how consciousness works as the "Cartesian theatre."[3] Dennett sees a parallel between this self-based perspective on consciousness and Descartes's idea of a soul that intrinsically possesses the power to make things conscious. In either case, whether it is a disembodied divine soul or a mechanism in the brain, this way of thinking passes the buck of explaining consciousness onto a theoretical homunculus.

This term originates in sixteenth-century alchemy and refers to a miniature, humanlike being that some tried to create though alchemical procedures. The physician Paracelsus suggested a strange method for creating a homunculus in his work *De natura rerum* (1537).[4] He wrote,

That the sperm of a man be putrefied by itself in a sealed cucurbit for forty days with the highest degree of putrefaction of the venter equinus [warm fermenting horse dung], or at least so long that it comes to life and moves itself, and stirs, which is easily observed. After this time, it will look somewhat like a man, but transparent, without a body. If, after this, it be fed wisely with the Arcanum of human blood, and be nourished for up to forty weeks, and be kept in the even heat of the horse's womb, a living human child grows therefrom, with all its members like another child, which is born of a woman, but much smaller.[5]

In cognitive science, researchers seek insight into how certain mental functions are performed. If I want to understand how seeing works, then suggesting that there is a brain process that simply looks and sees what is appearing on the retina is no explanation at all; it simply delays explaining how seeing works. Such an "explanation" has come to be known as a homunculus in cognitive science, as it is as if one is simply suggesting there is a brain process that functions as a miniature version of a person inside us, something that simply possesses the very capacity that we are trying to explain. The core problem with this approach is that it does not explain how the capacity in question is implemented in the brain. We could in turn ask how that homunculus has that capacity, such as the ability to make something conscious. Perhaps it, too, has a homunculus inside it, and we enter an infinite regress, with the question of how things become conscious never being addressed.

The conscious self does not seem a promising avenue for explaining experience. Not only are there philosophical problems with explaining the emergence of consciousness with reference to a self, but also the self may not even exist in the way we think it does. We typically think of the self as the solid and stable something that we are, the one who has experiences, makes decisions, and moves the body. It turns out that no such self exists. What we take to be the self is a construct, a perception, an image in the mind.

Take a moment to consider what your sense of self is. We know that we are physical, evolved organisms, so perhaps we will be on the most solid ground if we start with the body. If you think you are your body, then consider if you would still be you if you had a prosthetic limb. It seems reasonable to say that you still would be you, but add a prosthetic heart, mechanical lungs, a robot stomach, and keep going until you have replaced all your bodily organs with a mechanical equivalent. Are you still you? Is there a magical dividing line where you stop being you? It would seem that the body isn't the same thing as the self.

We typically feel like we exist apart from the world and even the body, commonly in our head, behind our eyes. Studies have been conducted that actually measure this: When an object is held the same distance from one's foot and from one's face, participants typically report it as being closer to them when it is near the head, as opposed to other parts of the body.[6]

In both cases the object can be equally close to the body, indicating that the body is not what people are referring to when they conceive of a self. Similarly, when we ask what happens to us after the body dies, the question reveals the intuition that you are something different from your body.

If you are not your body, then are you your mind? Our mental activity consists of a constant flow of appearances. Many meditative traditions focus on observing this impermanence and, by doing so, realizing that there is nothing stable that could be called a self amid the torrent of ever-changing appearances. The enlightenment philosopher David Hume came to a similar conclusion about what we take to be a self, deeming it a mere bundle of perceptions.

What if we were to look for a stable self in the physical brain? When we look for the self, difficulties arise because it isn't a solid thing that exists in the world. The self is a complex mental construct. Looking for the self in the brain is like looking for a video game character in the microchips of a computer. While we do not have a self that objectively exists, what an organism like yourself does have is the ability to model its own bodily activity, to store memories, to make plans for the future, and to weave all that together into a complex model that allows it to function in the world. At the center of this model is a division between the organism and its environment, and this is how the image of a self is made. It starts out as linked to the body because it exists in order to help the organism get around and survive, but it isn't the same thing. It is a concept, symbol, or image that allows the organism to discern what is the outside world and what is not. A self is a way of being rather than a thing. It is something that is performed, like a dance, more of a verb than a noun. When an organism is in the process of "selfing," it is functioning on the basis of an assumed separation from what it considers outside. When we stop selfing, we discover that the boundary is conceptual; in reality, we are not truly separate from the rest of existence.

We can readily see this sense of separateness to be an illusion when we consider the physical evolution of the universe and the emergence of life from chemical processes. While we feel separate from the world around us, we constantly take in food, water, and air to survive; there is a constant flow of energy through us. In technical jargon, we are open thermodynamic systems. It would be impossible for us to survive in a vacuum without the surrounding environment. In a very real sense, the environment is part of us: We couldn't exist without it. We see that scientifically we are not actually separate from the world. This sense of separation and the idea of a self that is set apart from the universe really is a psychological trick, not a fact of nature. We typically still carry this folk-psychological assumption of selfhood, however, and it pervades thinking at all levels in our culture, even in science and philosophy, where it proves a major barrier to thinking clearly about consciousness.

CONSTRUCTING THE SELF

How does the brain construct the illusion of the self? I explain in the last chapter the way in which perception can be thought of as a controlled hallucination, where best guesses about what is going on in the world lead to predictions about what inputs would be detected if such guesses were correct and are then adjusted based on actual sensory input. A guess can also be thought of as a hypothesis or a belief that things are a certain way. When perceiving anything in particular, the brain's best overall guess, or its big-picture belief, can be broken down into multiple sub-beliefs, each about what will occur at a different resolution. If I see what I believe to be a cow with its back half hidden behind a tree but the cow is walking forward, then my brain might have a zoomed-out, big-picture belief that this is what is going on in the scene. Such big-picture beliefs are thought to exist at the top of the brain's perceptual processing hierarchy, in the areas furthest from the sensory input. At the next level down, there are beliefs about what is predicted to happen in more specific areas of the scene if this big-picture belief is correct; for example, the tree should stay still, while the cow may continue to move. This process continues all the way down to the bottom of the hierarchy, where very specific sensory inputs are predicted; for example, the horizontal edges at the top and bottom of the cow's body should stay horizontal until its back legs become visible. The result of this hierarchical stack of multiscale beliefs is that the brain can operate in the most energy-efficient way possible. If the guess is proved correct by the sensory input that it receives, then the brain doesn't have to do anything. It only has to exert further energy if an error in the prediction occurs.

The beliefs we form typically reflect useful patterns that we have detected in reality. At the top of the hierarchy, we have our highly abstract beliefs, images of complex patterns we have extracted from our experiences. The self is an image of such a pattern. We may feel as if our bodies are separate from the world around us, but in reality, they are instead like whirlpools in a body of water. We constantly breathe, we drink water, we eat food—all the atoms of our body come from the world beyond us and are in a constant state of turnover. In fact, a decade after you read these words, every atom in your body will have been entirely replaced. We are a process occurring in the world, fully part of it. Our bodies are thermodynamic ripples in the greater world, yet because of the pattern hanging together over time, we abstract the belief in a body separate from the world around it. Even though our bodies are made of the food we eat and couldn't run without the oxygen we breathe, we still typically perceive the body as a separate thing from the rest of the world. At an even more complex level of abstraction, you might develop beliefs about what kind of person you are based on your previous behavior. The result is the conjuring of an apparition, a seemingly solid image of a stable and reliable entity where only patterns truly exist.

Where in the brain does this conjuring of the self occur? While it takes the whole of the brain, as well as the whole of the body and the world beyond, to give the impression of being a self, the subject who experiences experience, neuroscientists have identified a particular constellation of brain areas as being particularly important for this process: the default mode network (DMN; see figure 3.1). The DMN sits at the top of the belief hierarchy in the brain. It has access to emotional and autobiographical information, as well as the nervous system's decision-making apparatus, and leverages these neural systems to weave the belief in a self.

You may have seen pictures of the human brain with blobs of color on them illustrating which areas scientists have found to be active in particular situations. The MRI technology that allows neuroscientists to peer beyond the skull in this way is only a few decades old. In the early days of this research, scientists would compare neural activity during a particular task with a measurement of baseline activity at rest. For example, if you wanted to see which areas of the brain are involved in reading, then you might conduct a scan while a subject reads and then again while they are not reading. Differences between the two scans that are large enough not to be attributable to chance are visualized as areas of color, and by this method, we can identify which brain areas are recruited by which processes.

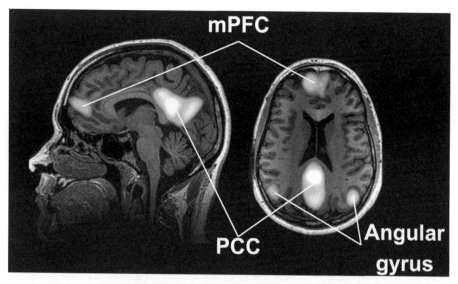

Figure 3.1. Brain scan showing the default mode network (DMN). The gray-white blobs show the key brain areas that make up the DMN. The left image shows a vertical cross-section of the brain from front to back, with the inside of one hemisphere and two major areas of the DMN that lie in the middle of the brain: the medial prefrontal cortex (mPFC) at the front of the brain and the posterior cingulate cortex (PCC) at the back. These two areas can be seen from another perspective in the image on the right, a horizonal cross-section of the brain viewed from above. The two other areas highlighted here by the gray-white blobs show an area of the DMN located on the outer surface of the brain: the angular gyrus. *Adapted from Wikimedia Commons. Original image from J. Graner, T. R. Oakes, L. M. French, and G. Riedy, "Functional MRI in the Investigation of Blast-Related Traumatic Brain Injury,"* Frontiers in Neurology *4 (2013).*

One day, some scientists had an interesting idea. When participants are at baseline, when simply lying in the MRI scanner without doing a particular task, they are not brain dead. When at rest, we still experience the world and ourselves in it; the brain is still active. The researchers wondered what kind of neural activity occurs when we're in our "default mode" of not engaging in any particular task. What potentially interesting patterns of brain activity was the field dismissing as mere "neutral baseline"? They analyzed the activity of the brain at rest and found a network of brain areas that were routinely active together in these states, when we are simply left alone and our minds begin to wander.[7]

The DMN can be thought of as underpinning this process of mind wandering. Imagine you are sitting on a bus by yourself, staring out the window. You're not engaging in a particular task, and your mind will most likely wander, from recalling memories to visualizing upcoming plans. Despite how effortless it may feel, mind wandering is a complex mental feat that involves simulating an image of yourself and projecting it into past and future circumstances. This highly sophisticated process is underpinned by areas along the midline of the brain where the two hemispheres meet, such as the medial prefrontal cortex (mPFC) and posterior cingulate cortex (PCC). These are the core brain areas that comprise the DMN, and they are also the areas that are dampened during mystical experience.[8]

Let's recap: In order to navigate in the world effectively and in an energy-efficient manner, the brain constructs a series of increasingly complex abstract beliefs about things that exist in the world. It divides the messy, continuous reality around us into distinct concepts. The most complex beliefs that the brain constructs include the belief in a self, an image of the organism that it positions in relation to the objects that it perceives. We are left with an image of what we are that doesn't correspond to a concrete reality; it is an apparition in the mind. In his book *Being No One*, philosopher and cognitive scientist Tomas Metzinger proposes a detailed theory of how it is that the self is constructed. He refers to the constructed self as an "ego tunnel," a distorted perspective on reality that allows us to focus on certain information that feels personally relevant, at the expense of much else.[9]

Let's end this analysis of the self with an experiment you can try yourself. Feel a sensation that is distant from your head—on your foot, for example. Now, touch yourself on your cheek. The sense of distance between "you" and your foot should be palpable, but interrogate how it is that you experience this distance. In reality, the sensation on your cheek and your foot are equally close to you from the perspective of your body as a whole. They are also equally close from the perspective of consciousness, as they both appear equally in awareness. Despite this, the sensation on your cheek *feels* closer to you. This is because your head is the center of your frame of reference.

Why should this be so? You might think it is because your brain is there, but that is not the primary reason. The main reason that you feel like you are located in your head is that this is where your major sense organs are located. Your brain is only located there to reduce the distance that sensory signals need to travel. If you could move your eyes, ears, mouth, and nose onto your right hand (excuse the grotesque image), then your brain might update its reference frame, and you might experience your self as existing in your right hand. In this way we can see that the self isn't the solid thing we often take it to be. It is a flexible construction.

DECONSTRUCTING THE SELF

Now that I have shown how the brain weaves our sense of self, it may make more sense that a chemical compound like psilocybin can produce a self-transcending mystical experience, but how does this happen precisely? When an individual ingests a psychedelic substance, the molecules move into the bloodstream and then into the brain, where they dock on specific cells. These brain cells, known as neurons, signal to each other electrically, but they pass on these electrical messages by releasing certain chemicals. Specific neurons associated with certain functions respond to different chemicals. The chemicals dock in specialized receptors on the receiving neuron and produce changes inside it. This is why drugs can work at all; drugs are chemicals that can interact with these receptors and imitate the activity of chemicals that already exist in the nervous system.

Psychedelics activate specific receptors that usually respond to serotonin produced within the brain. Neurons in the brain areas that comprise the DMN possess these receptors, and when psychedelics dock on them, they alter the activity of these cells, producing an overall suppression of their electrical signaling. This temporarily impairs their ability to function and, at high enough doses, stops them from successfully weaving the illusion of being a separate self, or any other conceptual divisions for that matter.

One model of how psychedelics act on the DMN to dissolve the self comes from two prominent neuroscientists, Karl Friston and Robin Carhart-Harris. Friston is arguably the most influential neuroscientist alive today, and he happens to be a former colleague of mine from University College London, where I conducted neuroscientific research for a number of years. (I present more from Professor Friston and his free-energy principle later in this book.) Carhart-Harris is one of the leading researchers in the field of psychedelics. These eminent scientists collaborated on a model of how psychedelics act in the brain called relaxed beliefs under psychedelics, or REBUS.[10]

The REBUS model is based on the hierarchical view of the brain described earlier. The DMN sits at the top of this hierarchy and is responsible for forming the highest-level beliefs that we hold. These beliefs are our most abstract and general beliefs, and they are simultaneously the ones that are most unshakable, such as beliefs about our ourselves and our world. These may seem like fundamental beliefs rather than high-level, abstract ones, and they are in fact both. They are extracted from the vast range of experiences you have had in your life and are highly constructed, unlike the belief in the yellowness of a banana in front of you, which is less constructed—although it is still a construction. While being abstract, they are the beliefs we lean on most heavily to navigate in the world, precisely because they are very general and work across a wide range of situations. Believing yourself to be a good person is a more useful belief in multiple situations than believing bananas are yellow, the latter being too specific and concrete to be helpful.

There is another sense in which these beliefs, despite sitting at the top of the hierarchy, are the most foundational. As explored previously, the areas at the top of the hierarchy, like the DMN, send predictions about what is expected to occur down to the lower areas of the hierarchy, based on the content of the high-level beliefs. If something occurs that contradicts these beliefs, the system can either explain the occurrence away, or if that is not possible, the beliefs can be updated to come in line with the new evidence. For example, someone who holds a negative belief about themselves as a person dismisses a small compliment, saying that the other person is just being polite and doesn't really mean it. Even though it may be emotionally more painful than accepting the compliment, this narrative allows the beliefs they use to navigate in the world to stay intact. These faculties evolved for our survival and not necessarily for our well-being, and even coping mechanisms that cause us pain can still serve to keep us alive. When faced with challenging circumstances, it is the belief that works for our survival that is kept. Now if this person undergoes psychedelic therapy, they may find that these negative beliefs about themselves soften, allowing evidence that they may not be so bad after all to be taken onboard.

The DMN also happens to contain a large amount of serotonin 2A receptors, the kind of receptor that is activated by psychedelics. According to the REBUS model, psychedelics disrupt the ability of the DMN to construct these high-level beliefs, making it more difficult for the evidence that contradicts them to be dismissed. This creates a powerful opportunity for these beliefs to be updated and be brought in line with what is the case in the present instead of leaving the individual to be haunted by negative beliefs formed in response to challenging or traumatic events that may have ended a long time ago.

How does this relate to the mystical experience? The belief that you are set apart from the rest of reality allows you to navigate the world in a way that benefits the organism. In reality, however, no fundamental separation exists. Under the influence of a psychedelic, this high-level belief in separation can fall away, leaving us faced with the evidence of the fundamental wholeness of existence. Without these beliefs clouding our perceptions, everything simply is, and nothing is truly divided from anything else. Anything else we can say about existence is just a story, a belief.

Can this approach be generalized to include nonpsychedelic-induced mystical experiences? A friend and colleague of mine, Shamil Chandaria, was a collaborator on the REBUS model and is an expert on how such altered states relate to the brain. Shamil is an expert meditator and has mapped this approach of thinking about the brain onto a system of meditative states called the *jhānas*. *Jhāna* is the word for *meditation* in Pāli, the language of the Buddha and of the oldest Buddhist texts. This word became *dhyāna* in Sanskrit, *chánnà* or *chan* once it migrated to China, and finally *chan* became *zen* in Japan.

There are eight *jhānas*, and as the meditator passes through each one, they deconstruct a belief they were holding about themselves and the world, until they get down to the subtlest beliefs imaginable. The process begins with absorbing one's attention in pleasurable sensations in the body. After a period of absorption in these sensations, the meditator feels the solidity of the map of the body dissolve. Once the physical sensations die down, a mental experience of well-being is revealed. Eventually, this, too, subsides, revealing a calmer state of complete, satisfied contentment. Beyond this state there is a state of even greater calm, of pure stillness and peace. This may seem like the final stop, but this is only halfway through the levels.

Beyond the stillness is a sense of totally unconfined spaciousness, where there is no sense of the body at all. Then, even the sense of space falls away, leaving only consciousness or awareness. Awareness here is revealed to be the backdrop against which everything in experience arises, no matter whether we typically label the arisings as internal or external. Shamil has suggested that these stages correspond to the deconstruction of increasingly subtle high-level beliefs in the brain's hierarchical system, until we see that what we experience as existing in here in the self and out there in the world are both actually appearances in consciousness.[11]

AWARENESS WITHOUT THE SELF

Who says words with my mouth?—RUMI[12]

We have explored how, in everyday experience, it typically feels like "I" am conscious and that "I" exist somewhere in my head, looking out of my eyes at the world around me. What would happen if "I" was not there? If it is the self that has experiences, then presumably experience would stop, as there would be no one there to experience it. The mystical experience reveals that this is not the case. In such states, the sense of self disappears or is felt to be secondary to a unifying awareness, yet surprisingly, experience does not stop. Instead, things continue to appear. Appear where? In the spacious, tranquil space of awareness. It is not the psychological sense of a separate self that is conscious; the psychological self is rather an appearance in an impersonal awareness. It is this foundational feature of consciousness, the illumination of experience, that science first needs to account for, what Metzinger refers to as "pure consciousness."[13] We should not be looking for the emergence of consciousness in the arising of a conscious self, as many prominent theorists do. We should be looking for the emergence of consciousness in the emergence of awareness and qualitative experience itself.

One way to understand the mystical state is as a figure-ground reversal in the prominence of consciousness and thought. Here I use *thought* to refer to the conceptual mode of consciousness. We are typically lost in our thoughts, seeing them not as things arising in the mind but instead as referring to very real events that occur somewhere else, somewhere that is not here. In the mystical state, thoughts can be seen for what they are: mental forms arising in experience. The impression that the content of thoughts is as real as whatever is appearing now is punctured. Thoughts no longer conjure an imaginary world that lies outside of the present moment. The present moment engulfs them, and they are seen for what they are: imaginings that are part of experience in the here and now.

With this, the impression that past and future are as real as what is occurring now is also dispelled. Consciousness is felt to be eternal, in the sense that it is does not relate to the paradigm of psychological time, which is based in imagined past and future. There is a finality to this state, a sense that there is nowhere to go and nothing to do, like coming up against a brick wall. Rather than feeling frustrating, however, this is an experience of gigantic relief, of feeling that it is okay to stop extruding oneself through time with all the stress for success and failure that that involves, and instead we find that it is possible to come to rest, to feel completely at home in existence.

With consciousness now in the foreground and thought in the background, consciousness is seen to have certain attributes that went previously unnoticed. Before such an experience, mental activity gives the impression that the past, the future, the three-dimensional environment around you consisting of different objects, your anxieties and concerns, and the burden of ensuring that your life goes optimally all have some kind of objective existence. In the mystical state, these are all seen to be appearances in consciousness. We can call these appearances the conscious content. Common to all this content, however, is the fact that it is conscious, it is being experienced, and it is arising in awareness. We can reserve the term *awareness* for this aspect of consciousness. In the mystical state, awareness is seen to be formless, in that the forms that arise in consciousness and make up the conscious content all appear in the same awareness that does not possess any form itself. You can experiment with this now by repeating any word in your head and seeing if you can notice the space of awareness within which this inner sound arises and disappears.

Perhaps most surprising of all, awareness itself is completely untouched by whatever arises in it. The analogy of the mirror is widely used to describe this aspect of the mind. In the same way that the reflections that arise in the mirror do nothing to disturb its reflective nature, whatever arises in experience, no matter how joyful or unpleasant, leaves the aware essence of consciousness completely unperturbed. No matter what appears, the mirror is unchanged. It is precisely because the mirror is empty of content that it has the reflective power to allow contents to arise in it. In the same way, the formless nature of awareness allows it to conform to the content that arises in the mind, like waves arising in a still lake.

Psychedelics can be used to powerfully alter the content of the mind. Even at very high doses, however, the formless aware core of consciousness remains unchanged. This can be a very surprising thing to discover, given the seeming dependence of experience on the brain and the physical organism. This gives some the feeling that what they are at their core is eternal and will be unaffected by the death of the body. It also leads some to conclude that awareness is at the core of existence, the stance of idealism. What's more, the appearances of both self and world appear to arise in the same formless awareness, giving the impression that the world is not actually divided into subject and object. This mode of consciousness is often referred to by researchers as "nondual awareness," as there is no duality between subject and object in this state.

The aim of this chapter and the previous chapter is to show that the beliefs in subject and object are just that—beliefs; they are mental constructs. What's more, they are mental constructs that block our understanding of consciousness. If these beliefs do not actually reflect our circumstance, then what is really going on? I next turn our attention to the most treacherous intellectual terrain of all: the very nature of reality.

FOUR

THE GOD WHO WASN'T THERE

Beyond Substance

The Tao is empty; when utilized, it is not filled up
So deep! It seems to be the source of all things. . . .
Is it not like a bellows?
Empty, and yet never exhausted
—LAO TZU[1]

CHEMICAL MYSTICISM

I sat on the floor of my houseboat as the sunrise reflected on the water of the London canal, the sound of chirping coots and moorhens drifted in through the windows. Lying in front of me were fifty-one grams of freshly picked psilocybin mushrooms. At the time I was living a nomadic lifestyle aboard this houseboat, the *Serendipity*. I spent the year cruising the nature-filled waterways of London, a refuge of greenery amid the concrete maze that I navigated on my way to the lab each morning. In my working life, I'd pored over the studies from my colleagues and understood the neurochemistry of what was about to happen. No amount of reading, however, could answer the question of whether these drug-induced mystical experiences are really the same thing as spontaneously occurring mystical experiences. As a neuroscientist who had had such an experience early in life, I was well placed to investigate this question firsthand, and when it comes to the inherently private phenomenon of consciousness, there are some instances where the only way forward is self-experimentation.

There's a thought experiment in the philosophy of consciousness that involves a neuroscientist called Mary who lives her entire life in a black-and-white room but knows everything there is to know about the neuroscience of color perception.[2] Somehow, despite all this knowledge, she learns something new when she exits the room and sees the blue of the sky for the first time. I now felt like Mary, about to open the door out of her tiny black-and-white world. I opened wide and plunged into the unknown.

Sometime after I had ingested the mushrooms, the world looked very different. As I looked around me, the boundaries between objects became compromised. Everything blended with everything else, and in a moment of exquisite relief, the subjective boundary between myself and the world suddenly dissolved. All that was left was a beautiful, shifting mosaic of awe-inspiring geometric patterns arising in consciousness. There was no center, no self looking at the experience. Everything arose equally in awareness. I was left for hours in a state of raw feeling without concepts. Reality billowed forth relentlessly. My mind repeatedly attempted to grasp onto some concept, any comforting abstraction that might temporarily give me some respite from the excruciating pleasure of pure being. Moments later, the folly of conceptualizing would be subsumed by the overwhelmingly obvious fact of pure existence. The concepts were arising in and as existence and so were powerless to obscure it. How did I ever overlook this fact? How did I ever get anything done when existence was so rapturously awe inspiring, so much more real than the ephemeral world of thought? Would I ever be able to overlook this fact again and live through abstractions as I had done before?

My mind inevitably cobbled itself back together as the chemicals left my system and my brain began to function normally again. While it had its own particular character, this experience led me to the same insights as the spontaneous mystical experience of my youth, again showing separation, self, and substance to all be mental constructs. It was a confrontation with the nonconceptual nature of reality and our experience of it. We really exist, and existence is unnamable—it is beyond what we think about it. It was an experience of being a natural phenomenon, an elemental force like a surging geyser or a tornado, something vastly more real and powerful than the abstract mental image we typically have of ourselves. To feel that this is what we truly are, underneath the stories we tell about ourselves, is to vividly experience what science tells us: that we are an evolved natural phenomenon in just the same way as whales, flowers, or bacteria. Such experiences leave us with the impression that consciousness, far from being ephemeral or epiphenomenal, is something like a powerful, elemental force deeply tied up with what it is to be in the world.

At the peak of the experience, there was no separation between any appearance in consciousness, no sense of self apart from the appearances, and no sense of a solid ground underlying existence. Yet, in the midst of the relentless flow, there seemed to be a stable stillness. If you've come this far, then you've tolerated the possibility that objects may not have their own independent existence and even the idea that you might not exist as a separate self. Now let's truly pull the rug from under your feet, as I question the nature of the very ground of existence.

THE NATURE OF REALITY: SUBSTANCE AND METAPHYSICS

If you put these two fundamental Buddhist ideas together—the idea of not-self and the idea of emptiness—you have a radical proposition: neither the world inside you nor the world outside you is anything like it seems.—ROBERT WRIGHT[3]

Our evolved minds find the concept of substance very useful. If I show you a toy castle and tell you it is made of Lego bricks, then you will understand the relationship between the bricks and the overall structure: The Lego bricks are the substance out of which the castle is made. This way of thinking works well for things like tables made of wood or knives made of metal, but we have a habit of believing that this way of thinking should generalize to the very fabric of reality. It is possible that our quaint primate modes of thought are not suitable for conceptualizing something so fundamental. We must remember that we are evolved animals that did not develop to see the essential truths of existence but instead for survival. Those of us with full-color vision see greens and reds because our species evolved to pick out fruits from leafy green backgrounds, but we do not see ultraviolet or infrared light as some other animals do because that wasn't useful for survival in our evolutionary niche. There is no reason to think that our minds should be able to intuitively grasp the operation of existence itself. Seen from this perspective, it would be odd if reality were indeed built of a stable substance, whether matter or mind. This would also raise the difficult issue of how this solid stuff came into existence in the first place, of what gives it its substantial nature.

We typically think of substances as possessing solidity or stability, so let's examine the experience of solidity in physical objects to see what is really going on. Whatever you are sitting or standing on could be said to exist mainly of empty space, but due to the laws of physics, you do not slip through the object. Not only are the object and your body mainly empty space, but also the particles that make them up are in constant motion. So we can see that at the microscale there is no solidity and stability, but at the macroscale, the scale at which you interact with the object, there is a relationship where you cannot easily pass through it. The brain represents this lack of the possibility of movement as solidity, giving the feeling of a solid material substance existing. If you were the size of a quark, perhaps the object would not be perceived as solid. We project our impression of solidity onto the object itself, which gives us the impression of it actually possessing that quality. In reality, we are making something that is entirely contingent on our relationship with the object into something that is an objective feature of it, and this is a mistake. In this way we give ourselves the impression of a world that is genuinely made of a solid material substance rather than one that only feels that way, due to the ways in which we seek out and perceive stability.

What could reality be at the fundamental level, if not a substance? Questioning the nature of reality itself takes us into a branch of philosophy called metaphysics. While physics is typically conceived of as the most fundamental science, metaphysics takes us even deeper into the core of existence. Whereas physics is concerned with the study of the natural world and the laws that govern its behavior, metaphysics is concerned with the most fundamental questions about reality, including the nature of existence, causality, time, and space, as well as the relationship between matter and mind.

Metaphysics has become largely taboo in Western academic institutions. This is understandable because claims regarding the fundamental nature of reality are typically untestable and highly speculative. Rather than succeeding in ridding us from metaphysics, however, this taboo serves to let a naive reductionist materialism operate unconsciously in the background. Despite the inherent limits of the metaphysical project, we must engage this terrain if we are to reconcile matter and mind. If we leave our commonsense folk-psychological notions of self, separation, and substance uninterrogated, I believe we will not be able to make sense of consciousness. We must do metaphysics but with humility and an awareness of its speculative nature.

THE PERENNIAL PHILOSOPHY

One reason to take the core insights of the mystical experience seriously is that they occur in many different religious traditions around the world. The idea of the universality of these insights is known as the perennial philosophy, a term popularized by the famed writer Aldous Huxley.[4] According to this perspective, the concept-dismantling mystical experience lies at the heart of the world's religions. With one's mental constructs suspended, how is one to articulate the vision of bare nonconceptual existence that is beheld in that state? It is plausible that this is where the term *God* originated in monotheistic religions, to refer not to a supernatural authoritarian creator but as a pointer to that which is so ultimate that it transcends concepts. This would be a metaphysical ground that could be compatible with science, as there is nothing supernatural about the idea of the world existing in itself before we lay our concepts over it.

The mystical core of many religions is still very much alive. In the Vedic tradition of Hinduism, the entire universe is seen as a manifestation of a single ultimate principle, known as Brahman. The goal of Vedic spirituality is to experience the unity of the essence of the individual, known as Atman, and the whole of existence, Brahman. In some schools of Buddhism, the ultimate reality is the primordial enlightened Buddha nature, which pervades all phenomena. The founder of Sikhism, Guru Nanak, also points to this idea of God as the experience of the unity of existence that can be "attained" by the religious practitioner: "There is but One God. His name is Truth. . . . If there is one God, then there is only His way to attain Him. . . . Worship not him who is born only to die, but Him who is eternal and is contained in the whole universe."[5]

The Abrahamic religions, too, have their mystical elements. In these traditions, God is typically seen as the foundational principle of existence, from whom all things originate and to whom all things return. Judaism, the foundational Abrahamic religion, has Kabbalah, a mystical tradition that seeks to directly experience the nature of God and the universe through the study of sacred texts and the use of meditation, contemplation, and ritual. Islam, the most recent Abrahamic religion, has Sufism, a mystical body of practices that also aims at connection with the divine.

Christianity has its own mystical heritage in the form of Gnosticism, a philosophical movement that emerged in the early Christian era that held that direct experiential knowledge, or gnosis, of the divine is the key to salvation. Mystical traditions that emphasized the direct experience of God through prayer, meditation, and contemplation continued into the Middle Ages in the work of figures such as Meister Eckhart, Saint John of the Cross, Julian of Norwich, and Saint Teresa of Avila. In these traditions, the subjective realization of nonseparation from God is associated with the experiential attainment of paradise or heaven. "The Kingdom of God is within you," as Jesus is said to have put it.

These traditions were influenced by a philosophical and religious movement that emerged in the third century CE in the Hellenistic world called Neoplatonism. At its core Neoplatonism posits that the ultimate reality is a single, ineffable source of all being, known as the One. The material world that we experience is seen as a reflection or emanation of this divine reality, and the goal of Neoplatonic spirituality is to unite with the One through contemplation and ascetic practices and thereby achieve a state of ultimate happiness and fulfillment.

While the mystical experience of the individual may lie at the core of many religious traditions, a countervailing trend of gatekeeping also exists. In Christianity, mystical sects that sought to bring these experiences to everyday people were aggressively stamped out by the centralized church hierarchy that sought to maintain power for themselves. In fact, the Inquisition began not as an attack on Jews and Muslims but on Gnostic Christians. If you can experience union with God for yourself as a Gnostic, then you may be less likely to respect the men with the fanciful hats who would rather you believe God is a supernatural being who lives somewhere very far from your personal experience. The religious traditions that hoarded power for personal gain and kept the average person alienated from direct religious experience are consistently associated with such notions of a transcendent God existing somewhere else and not being imminent in one's direct experience of existence.

It is this version of Christianity in which we are alienated from God that the philosophers who founded modern science were adherents of, and science is still infected with some of the assumptions of this worldview. Descartes proposed consciousness as originating in a divine soul that is imbued with its powers by a transcendent God who exists outside the natural world. As a result, the world was seen as a lowly mechanism. Why did Descartes associate consciousness with religious experience of the divine? In atheistic circles we are typically given the impression that these thinkers of the past were religious because they were simply ignorant and culturally brainwashed by the ideology of the time. There is another possibility, however, and that is that direct experience of nonseparation is an important piece of data to reckon with when constructing a worldview. Perhaps we must do what many cultures have done in the past and once more reconcile the inner and outer ways of knowing into a coherent nonsupernatural picture of reality.

SPINOZA'S INFINITE GOD

God is the indwelling and not the transient cause of all things.—BARUCH SPINOZA[6]

Baruch Spinoza was an admirer of Descartes but offered a very different picture of reality—one that could save us from Descartes's dualistic legacy. Spinoza was a seventeenth-century philosopher whose work relies to an incredible degree on rationality. His magnum opus, *The Ethics*, consists almost entirely of statements that he combines with practically arithmetic precision. To some, he is history's first recorded atheist. Spinoza lived in a Portuguese Jewish community in the Netherlands that had been founded as a result of the Inquisition on the Iberian Peninsula. His synagogue in Amsterdam holds records showing that he was excommunicated for his beliefs as a young man. While the records do not explain the details of these heresies, it seems plausible that the young thinker was perhaps expounding on the ideas he would later write about in his philosophical work. What we do know is that Spinoza's "wrong opinions," as they are described in the excommunication edict, led to his community's elders voting unanimously to impose Herem, or excommunication, on Spinoza, stated in the strongest possible terms. The decree states,

By the verdict of the angels, and the judgment of the saints, we impose Herem, expel, execrate and curse Baruch Spinoza. . . . Adonái (Lord Yahweh) does not forgive him. May the wrath and anger of Adonái be unleashed against this man and throw upon him all the curses written in the Book of this Law. Adonái will erase his name under the heavens and expel him from all the tribes of Israel. . . . We warn that no one can communicate with him orally or in writing, that no one pays him any favors, that no one remains with him under the same roof or less than four yards away, that no one reads any text produced or written by him.[7]

It came as a great surprise to me that my eleventh-great-grandfather Jaacob Barzilay had signed that document in his role as a member of the Lords of the Ma'amad, the governing body of seven hakhamim, or wise Torah scholars, who oversaw the community. I knew that I had Portuguese Jewish ancestors in Amsterdam during that time and that the community numbered only in the hundreds, but as a fan of Spinoza, I did not expect one of my direct ancestors to be directly involved in his shunning.

So what were Spinoza's views that may have resulted in this expulsion? In *The Ethics*, Spinoza first turns his attention to the fundamental nature of reality to put his ethical system on maximally solid footing. Spinoza was a monist, like the idealists and the materialists. While the idealists argue that the one fundamental thing is mind and the materialists argue that it is matter, Spinoza proposed a third option. He set out a metaphysical picture that has come to be known as neutral monism. In this perspective the fundamental substance of reality is conceived to be something that is neither matter nor mind.

Spinoza argued with extreme rigor that whatever the fundamental substance of reality is, it must be self-caused; that is, it must rely on nothing other than itself for its existence. If I claimed that reality was made of tiny material building blocks, then it would be perfectly reasonable for you to ask what those blocks are themselves made of. The fundamental basis of reality can't be something of this kind, as it can't be considered fundamental if it is itself made up of something even more fundamental. To be truly fundamental is to exist in one's own right. Spinoza also argued that this substance can only exist as a unity, as there cannot be multiple separate fundamental substances underlying reality. He also claimed that it cannot be finite itself if it is to be conceived of as underlying all finite forms. The unlimited or infinite nature of this basic substance must also extend in time, making it eternal. Here we have a self-caused, unitive, infinite, eternal, fundamental ground through which all the drama of existence takes place.

This ground is neither matter nor mind, both of which are considered to arise within it. To Spinoza, matter and mind then are not really separate but are different "modes" through which this fundamental substance operates. We can think of mind and matter as two different types of organization that can occur within the whole that is reality itself. Spinoza sets out every aspect of his thinking behind these claims with a bone-dry rigor that can be exhilarating for a certain kind of logical mind and probably incredibly dull to others. It is like watching someone slowly putting in place an array of seemingly mundane structures only to realize that they are in the process of cornering and trapping the essence of existence.

What are we to call this single, infinite, eternal, self-causing substance that forms the basis of all of reality if it is deeper than both mind and matter? Spinoza opted for *God* or *Nature*, with *Nature* understood to be the totality of the natural world, the entirety of the cosmos, not just the green stuff outside our windows. This is not the personified God of the Bible, however. Spinoza argued for a nonsupernatural vision of God. For Spinoza, God and Nature were entirely the same. He sums this up in the Latin phrase *"Deus sive Natura"*—"God or Nature"—illustrating their total interchangeability in his eyes.

It was perhaps this equivalence between God and Nature that got him in such hot water with the religious authorities while in his youth. Perhaps they felt that this vision of God as Nature was effectively no God at all. Whichever word we choose, we can see here that, with this move, Spinoza was truly synthesizing science and religion. Despite his extreme rationality and arguable atheism, Spinoza was clearly familiar with the mystical sense of the unified nature of reality. For this reason, many call his position pantheism, the idea that what we call God is the same thing as the totality of existence. Albert Einstein is known to have said that he believed in "Spinoza's God," illustrating the compatibility of this perspective with serious scientific thinking. I typically see the word *God* as misleading and confusing given its dominant cultural associations with a transcendent supernatural being, but this picture makes me far more sympathetic to its use. If *God* refers to anything, surely it is that which is so fundamental and absolute that it is beyond all attempts to objectify and categorize it. The labels here are ultimately a matter of personal choice and the pragmatics of communication. What is important is that there is nothing supernatural in this worldview, yet it accommodates both mind and matter, as well as science and mystical experience.

REALITY AS A RELATIONAL PROCESS

While Spinoza uses the term *substance* to describe what God or Nature is, we can differentiate two ways in which this term can be used. One way is to conceive of a substance as a kind of material, something with solidity and stability. *Substance* can also be used to mean that which is most fundamental, that which grounds and underpins everything else in existence. This second definition is the way in which Spinoza uses *substance*, but confusion can arise given these two interpretations of the term. What we seem to be dealing with here is therefore a kind of insubstantial substance or groundless ground, something that underpins reality while not being objectifiable as something solid.

Spinoza's God has no finite form in itself. It is the nonfinite ground that has the ability to contain all finite forms. It is simultaneously empty of any fundamental nature, which allows it to contain every kind of nature possible. Spinoza's description of God appears to be similar to the Chinese concept of the Tao. The Tao, or "the Way," is the metaphysical ground of existence that underlies all phenomena but cannot be captured in words, according to Taoism. "The Tao that can be spoken is not the eternal Tao," state the opening lines of the Taoist classic the *Tao Te Ching*. Much like Spinoza's God, it is the nonfinite and inherently empty nature of the Tao that allows it to become everything in existence.

This is a vision of the ground of reality that avoids talk of substance. Instead, reality is seen as a process in the way that a river is a process. In the Taoist system, all forms arise out of this empty ground through a process of polarization, or mutual codependent arising. We saw this interdependent or relational vision of reality earlier, when we explored the concept of emptiness. Night and day, up and down, being and nonbeing—all are defined against each other, they could not exist without their opposite. This unity through duality is illustrated in the iconic yin-yang symbol of Taoism. All forms that arise in this empty ground, however, are impermanent, as there is no solid foundation within which they could stabilize themselves. In the Taoist view, reality is seen as a process rather than as being based in solid substance.

Here we have a process-relational view of the world as consisting of a flow of interdependent phenomena. A world away from China in Mesoamerica, the Nahua peoples, the most famous of whom are the Aztecs, developed a remarkably similar cosmology. They, too, conceptualized all of reality as the play of a fundamental process, which they called Teotl, rather than Tao. This Teotl is also seen as creating the structure of our world through mutually codefining polarities, as in Taoism.[8] It is not just Asian and Indigenous American religious traditions that present a view of the world as a relational process. The word *physics* comes from the ancient Greek word *physis*, meaning *nature* or *essential nature of all things*. This term itself comes from the word for *growing* or *becoming*, illustrating this nonstatic vision of our universe. The ancient Greek philosopher Heraclitus explicitly advocated for a process metaphysics, famously writing, "No one ever steps in the same river twice," and teaching that "everything flows" and that "all entities move and nothing remains still."[9] He also expounded on the same idea of structure existing through the relational unity of opposites, as in one of his aphorisms: "The road up and the road down are the same thing." He saw shifts between opposites as the basis for his cosmology of change: "Cold things grow hot, hot things grow cold, a moist thing withers, a parched thing is wetted."

What are we to make of these same process-relational views of the world popping up around the globe? One possibility is that, as with the perennial philosophy, they all arose out of an examination of direct experience. When we see substance to be a mental construct, we are left with a vision of the world as interdependent flow staring us in the face.

Further validation for the idea that we live in a process-based reality and the idea of substance being a mental construct can be seen in how scientific understanding has progressed over time. Before we develop genuine insight into the nature of a given phenomenon, we typically attribute its particular characteristics to some substance that seems intrinsically to possess these qualities. Before we began to understand the physical processes involved in life, living things were imagined to possess a vital substance. Before the process of combustion was understood, this phenomenon was attributed to an imagined substance called phlogiston. The same is true for our understanding of heat, which was originally thought to occur due to a substance called caloric. In the nineteenth century, it was thought that light was transmitted through a substance called luminiferous ether. Until relatively recently the divine soul was posited as the substance of the conscious mind.

One can see the history of science as a process of transitioning from substance-based explanations for phenomena to process-based explanations. This trajectory would indeed make sense if it were the case that reality really is a process. Belief in a substance that exists for no reason, such as a soul or a vital force that gives things life, is classic supernatural thinking. Science has systematically dismantled these substance-based explanations, and as I show in the next chapter, it is now doing so with matter itself. Consciousness is the final phenomenon that we must use science to understand in process terms.

SOMETHING FROM NOTHING

Seeing beyond concepts appears to present a vision of reality in which, rather than being grounded on a solid, stable substance, all forms arise through an interdependent process that emerges out of a nonfinite formless ground. This might sound quite out there, but I believe this way of thinking offers a truly nonsupernatural picture of reality. In our current scientific worldview, we simply must proclaim that certain fundamentals, such as space and time, exist for no reason. While we are at it, we might as well say a creator God exists. Starting with brute assertions of this kind is supernatural thinking, whereas in a naturalistic framework, everything must be explained and accounted for. We must start our explanations from nothing.

We can now sketch a speculative metaphysical picture of reality. Keep in mind that these ideas cannot be proven one way or the other; this story is presented here in the hope that it might give an intuitive feel for how the somewhat-radical claims that have been made so far might be valid. The logic goes like this: A naturalistic vision of reality must begin with nothing, as positing the existence of something with no reason, like God or absolute space-time, is supernatural thinking. This nothingness is our fundamental "ontological primitive"—the first thing that exists.

The idea of nothingness existing has been seen as an impossibility in much of Western thought since Parmenides, who argued that "whatever is is, and what is not cannot be." From this perspective, the existence of nothingness is seen as a logical contradiction. In much of Asian philosophical thought, however, this stumbling block is avoided due to a widespread appreciation that the territory of reality itself lies beyond our conceptual maps of it. As a result, many traditions, from Buddhism and Hinduism to Daoism and Neo-Confucianism, as well as modern Korean and Japanese philosophy, are perfectly comfortable posing the idea that reality is founded on an absolute nothingness that transcends the dichotomy of existence and nonexistence, being and nonbeing, or something (relative to nothing) and nothing (relative to something).[10] A related stance can be found in the work of French philosopher Jean-Paul Sartre, who wrote in his work *Being and Nothingness*, "Nothingness lies coiled in the heart of being."[11]

So we begin with truly nothing. What is the nature of true nothingness? Limits or boundaries cannot exist within true nothingness, so its nature is unbound, unconstrained, infinite. This is synonymous with pure potential, as in an infinite space, anything that *can* happen *will* happen. At the bottom of existence, according to this view, is pure, unbound potential, empty of any inherent being in and of itself. What kinds of events could happen in this space of empty potential? Given that it is nonfinite, anything that could possibly happen will happen. In this way, nonbeing evokes being; formlessness evokes form. Solid, stable forms would not be able to persist if they were grounded in nonbeing, but impermanent, insubstantial events or occurrences could arise in a relational manner. Why? Because relational things can be defined against one another. Up is not down, hot is not cold, and so on. It may be possible for a relational flux of this kind, with each aspect being defined against other aspects of the flux, to emerge in this potentiality. Reality would therefore consist of an eternal process of insubstantial events arising in a relational manner, an ephemeral cosmic house of cards in perpetual motion, with no solid foundation.

In such a framework, nonbeing evokes being, while being is dependent on nonbeing for its existence. This interplay acts as the furnace that continually forges existence. We could call this a kind of "dialectical monism," as *dialectic* refers to the interplay of opposites. In this picture formlessness, nothingness, or nonbeing can be said to play a fundamental role in the formation of form, somethingness, or being. This codependent relationship is captured in the Buddhist Heart Sutra, a core text for Zen Buddhists. "Form is emptiness, emptiness is form," the text enigmatically states.[12] Zen is deeply influenced by Taoism, whose core texts also allude to the generative power of absence. They point, for example, to the emptiness of a pot being the very feature that allows it to be filled. The introduction of the concept of zero fundamentally changed mathematics, having been previously viewed as suspect. Something analogous may be possible with the introduction of a fundamental role for formlessness in metaphysics.

This interplay of formlessness and form should not be conceived of as a process that happened a long time ago; in this view, it is happening even now—there is nothing that ever happens that is not this. Is it really possible for something to come from nothing, though? In physics there is the zero-energy universe hypothesis, which holds that the total positive energy in the universe, which takes the form of matter, is cancelled out by the total negative energy, which takes the form of gravity. As there is no net positive or negative energy in this view, such energy dynamics could have arisen from nothingness, with positive being defined against negative and vice versa. One proposal is that these dynamics could have arisen due to quantum fluctuations in a vacuum.[13] A vacuum is not true nothingness, however, and this hypothesis is part of physics, not metaphysics. We will never be able to scientifically demonstrate the truth or falsehood of these metaphysical ideas, but the zero-energy universe hypothesis illustrates the validity of this idea of mutual co-arising through polarization.

Everything in existence is temporary, but some things are clearly less temporary than other things. Why do we live in a universe with forms that persist if the structure of reality is based on insubstantiality and impermanence? Why isn't everything constantly dissolving? Here we can use the same statistical principle that underlies natural selection to make sense of the existence of forms. With natural selection, species are sculpted by evolution to take on the form that suitably mirrors their niche: Giraffes have long necks because of the heights of the trees where they evolved, the body of a fish is matched to the dynamics of the water in which it lives, and plants photosynthesize because of the energy that is available from the radiation of the sun. This sculpting only occurs because of the fact of death. Without the cutting back of unstable forms, stable forms would not be selected;

instead there would be an abundance of noise and chaos. We can think of the same thing as happening at the base of reality. The formless void is both the space out of which the events that constitute being arise and the space that continually strips back these events, leaving only the forms that exhibit some stability. We can conceive of this as a kind of Darwinian metaphysics.

This is a perspective where reality is an eternal, unending process. This means that reality would continue beyond the end of our own universe. This does appear to be feasible within our current understanding of physics, but these possibilities are naturally speculative, as they lie outside the horizon of the observable universe. One such possibility, known as cosmological natural selection, is that the singularity that we observe as a black hole from within our universe is seen as a big bang on the other side. If this is the case, then universes beyond ours could be born forever, with forms emerging out of the potent, formless void and dissolving back into it in an eternal dance.

This emphasis on formlessness tests the limits of our conceptual abilities, as it isn't actually a thing that can be objectified. The term *formlessness* is being used to point to intuitions of substancelessness, of solid substance being a mental construct. We could also call it the *insubstantial*, defining it in terms of what it is not: that is, a solid substance. In the place of substance, we have a process-relational metaphysics in which existence is conceived of as a single, ever-changing independent structure with no solid foundation. As was outlined at the start of this exploration, metaphysical claims should be held lightly, and setting aside the majority of these philosophical claims will be of no consequence when it comes to understanding the scientific ideas proposed in this book. A process-relational metaphysics in which the idea of substance is seen as a mental construct, however, is necessary to make sense of consciousness.

If we hold onto the idea that matter is a solid material substance, then we will not be able to make progress. We must instead appreciate, as most philosophers do, that physics studies patterns in reality at the microscale, not the nature of a substance that we ever actually come into direct contact with. You may be wondering at this point if in this analysis we are moving away from the scientific vision of the world in which all the phenomena of nature emerge out of the operation of the physical world and toward some form of idealism. This is not the case. I next explore the stance of idealism and the nature of matter and conclude that a philosophically nuanced nonreductive interpretation of physicalism, as opposed to naive materialism, is the most coherent of these perspectives.

NONDUAL NATURALISM

The last three chapters have been devoted to presenting a worldview based in naturalistic scientific and philosophical thinking combined with the experiential insights of the mystical experience into nonduality or nonseparation. I call the resulting perspective nondual naturalism. Nonduality is the recognition that reality is an indivisible whole and that perceived separations are constructed. Reality is nondual in that it is not fundamentally separated in two anywhere; it is a coherent whole. Naturalism is the position that the behavior of all that exists is consistent with what we know about the world through scientific observation. This arises out of a deeper commitment to having our understanding of the world cohere, to not tolerate any fundamental divisions.

It turns out that these two positions align perfectly, creating a synthesis of the scientific and spiritual worldviews. In this picture we have a universe that hangs together as a seamless whole and an attempt to map that reality in a similarly coherent manner. According to nondual naturalism, we are not separate from wider reality; we are entirely embedded within it. This gives credence to the claims of mystics to be able to know reality by feeling their union with it, as it is true that we always exist within the greater whole whether we notice it or not.

What makes this worldview different from other spiritual worldviews then? The commitment to naturalism, to scrutinizing our claims and checking their compatibility with the available scientific evidence is the key difference. As people open their minds to the possibility that there may be more to existence than they previously thought and subsequently find this open-mindedness validated, they can move further and further away from rigor. The wise balance is to be open yet sufficiently skeptical at the same time. As Professor Walter Kotschnig supposedly told his students, "Keep your minds open—but not so open that your brains fall out."

What does this worldview offer when it comes to explaining the presence of consciousness in our world? Nondual naturalism is deeply monistic and so does not accommodate incomplete, supernatural, or dualistic explanations for phenomena such as consciousness, seeing all perceived separations as constructed. It is naturalistic in that the explanations for the phenomena of nature are held to exist in nature itself and not in some supernatural elsewhere. We see that consciousness is not dependent on a self that has the power to introspect, and we see that the core of consciousness is an unchanging, formless awareness. It bursts the illusion that reality is made of solid material substance and replaces it with a process-relational metaphysics, in which all of reality consists of an interdependent flux. I show in the next chapter that this view is now being validated by physicists, as I consider what exactly matter is.

FIVE

WHAT IS MATTER?

An electron is a particular type of regularity that appears among measurements and observations that we make. It is more pattern than substance. It is order, but of the most clean and crystallized variety. Thus we arrive at a strange place. We break things down into smaller and smaller pieces, but then the pieces, when examined, are not there. Just the arrangements of them are. What then, are things, like the boat, or its sails, or your fingernails? What are they? If things are forms of forms of forms of forms, and if forms are order, and order is defined by us (who define macrostates) and by history (which actualizes them) and by the Universe (which undergirds the order), then those forms, it would seem, do not exist in and of themselves. They exist, it would appear, only as created by, and in relation to, us and the Universe. They are, the Buddha might say, emptiness.
—ANTHONY AGUIRRE[1]

A LEGO-BRICK UNIVERSE

When asking how consciousness relates to matter, it is easy to take for granted that we know what matter is. In the first few decades of the 1900s, scientists began to study matter in detail and were confronted with the fact that it seems to be nothing like we had imagined. Before these revelations, matter seemed straightforward. We look around us and see things that are made of smaller things—a wall made of bricks, a meal made of ingredients, a piece of fabric made of threads, a toy castle made of Legos. It doesn't take a huge leap of imagination to imagine those smaller things being made of even smaller things and so on and so on, until we end up at the smallest thing that something can be made of. This is the idea of the atom, which literally means *the indivisible*. The idea that indivisible particles make up the fabric of reality originated in both ancient Greece and ancient India. In India, the sixth-century-BC philosopher Kanada put forward this idea, and in the fifth century BC, the pre-Socratic Greek philosophers Leucippus and Democritus proposed the same idea.

In the 1600s and 1700s, Sir Isaac Newton developed his own atomic theory. This gave us the modern idea of the billiard-ball universe of particles moving through space over time as a result of being acted on by forces. Before the revolution in physics that took place in the early twentieth century, the worldview of science had been based on Newton's classical mechanics for more than two centuries prior. This branch of physics consisted of principles that were derived from the observation of large-scale objects, the kinds of objects we routinely interact with: tables and chairs, books and buildings. The apple that it is said to have hit Newton on the head is such an object. Newton would go on to be the founding father of this scientific discipline.

In the Newtonian worldview, the universe is a vast machine. Physical processes are taken to be what is real, with more complex phenomena like minds and societies being "epiphenomenal": that is, things that play no causal role in the world and are less real than the physical mechanisms underpinning them. In keeping with this, Newton advocated for Descartes's "method of analysis" or "reductionism," in which complex phenomena are understood by breaking them down into their component parts. If the parts are still somewhat complex, then you keep breaking them down until you find the ultimate underlying mechanism and, eventually, the basic indivisible unit that underpins the functioning of everything in existence.

In the Newtonian vision of existence, all reality is made of matter, a solid substance that is made up of material units called atoms. Here the mind-body problem rears its head again, the question of how to understand the nature of mind in this material paradigm—but I leave that aside for now. In this picture of reality, matter moves around through time on a stage of three-dimensional space. Space and time are separate from matter, and they are absolute. They simply exist with no prior cause and are themselves not made of any particular substance. This picture is also deterministic, with the movement of matter following laws of cause and effect. According to this view, if you know the initial positions and velocities of all the particles of matter in a system together, as well as the forces acting on them, then it is possible to accurately predict the behavior of the system with complete precision.

One of the most notable individuals who disagreed with Newton's concept of absolute space was the German philosopher and mathematician Gottfried Wilhelm Leibniz. Leibniz argued that space is a relational concept and that there is no such thing as absolute space. He believed that space is defined only by the relationships between objects and that there is no space outside these relationships. This view was in opposition to Newton's belief in the existence of an absolute and objective space that exists independently of the objects within it.

In the Newtonian system, it would be possible for us to stand back from any object we chose to investigate and examine it in the same way we might examine the operation of a machine. In this paradigm, the behavior of objects is assumed to rely only on the behavior of the particles of matter that make it up. As long as we don't disturb those particles, the functioning of the object should be unaffected by how we choose to observe it. The findings of physicists in the early twentieth century, however, brought this conception of reality into question. In his theory of special relativity, Albert Einstein showed that space is indeed relative, as Leibniz had argued centuries earlier. He even extended this relativity to time, introducing the concept of a unified space-time. This principle was applied at the cosmological scale but, at the same time, an equally significant revolution was occurring in our understanding of the smallest possible scale.

QUANTUM MECHANICS

In the early 1800s, evidence was found for the existence of what we know as atoms. By the end of the century, it had been discovered that the term *atom* could not have been a more inaccurate name for these particles. They were found to not be the smallest particle at all but to instead consist of electrons orbiting a nucleus of protons and neutrons. The study of subatomic phenomena in the first few decades of the 1900s brought into question the now millennia-old idea that the fabric of reality consists of small, indivisible, solid building blocks moving through space.

This subatomic scale is also known as the quantum scale, as these particles were discovered to exist in a number of distinct, quantized states rather than being able to flow continuously between these states. The behavior of these particles, or quanta, was found to violate the laws of classical mechanics that were thought to underlie the operation of the entire universe. Rather than simply moving through space like classical objects, an apple falling from a tree, for example, these quantum objects were found to be deeply influenced by the act of observing them.

In the famous double-slit experiment, a beam of light is shone at a barrier containing two slits. A striped vertical pattern appears on a screen behind the barrier, as would be expected if the light behaved as a wave. This is because the waves emanating from the two slits would cancel out in some places and sum together in others, creating what is called an interference pattern. When a particle detector is placed at one of the slits, the beam no longer appears to behave like a wave, and only two lines appear on the screen behind the barrier, one for each slit. This is the kind of behavior one would expect to see if the beam of light consisted of particles. The picture the new field of quantum mechanics was left with was one in which each photon in a beam of light exists as a wave function, a distribution of possible states that the photon could be observed in, up until the moment of observation, when it would be observed as a particle in a particular location. What form the quantum world took seemed to depend on our simply looking at it, a phenomenon known as the observer effect.

The fact that the wave function reflects a range of possibilities rather than a certain fixed future that depends on current conditions points to another important feature of quantum systems, known as indeterminacy. We think of large-scale objects as having their future determined by current conditions. According to Newton, if you know the location of an object in space, its velocity, and all the forces acting on it, then you can predict its future with certainty. In a quantum system, this is not the case. The precise location of the wave function in which a particle will appear is fundamentally unknowable; it simply appears in a way that is undetermined by current circumstances. All we can know is the probability of it appearing in any particular place—probabilities that make up the wave function.

Another strange feature of quantum systems is that they appear to be influenced by the circumstances in which a measurement is made, a phenomenon known as contextuality. You can measure a property of a particle known as its spin along different axes, but which axis you choose will affect the measurement you make. If you measure on the vertical axis, then you will find the particle spinning either up or down, but you will find a different result if you measure on a different axis. This behavior differs from classical physical systems, which we think of as possessing set properties that are not affected by our measurement of them. This would be like weighing a vase in its normal orientation, then weighing it again on its side, and finding that the way you weighed it changed the result of the measurement. If this were the case, could we say that the vase actually possessed a specific weight independent of our measuring it? Perhaps it would exist in a superposition of both weights until the moment a measurement occurred.

What's more, particles can become entangled so that the choice of measurement used for one particle influences the measured properties of the other particle, even over vast distances where a causal connection through space is impossible. This phenomenon is known as entanglement and is another mysterious feature of quantum systems. Finally, there is Werner Heisenberg's uncertainty principle, which states that, with certain pairs of properties of a particle, such as its position and momentum, both cannot be precisely measured simultaneously. When we gain information about one property, we lose information about the other property, again indicating that the properties of quantum systems do not exist in a pregiven state that we can simply observe without affecting the system. Any successful interpretation of the findings of quantum mechanics would have to make sense of each of these phenomena, including the observer effect, indeterminism, contextuality, entanglement, and uncertainty.

Two of the leading physicists of the field, Heisenberg and Niels Bohr, worked together in Copenhagen in the mid-1920s, and their interpretation of the many strange phenomena observed at the quantum level came to be known as the Copenhagen interpretation. In this view, quanta exist in a blurred superposition of all states they could possibly exist in, and the act of observation by a conscious observer causes this wave of possibility to collapse into a particle in a specific location.

A rival perspective, known as the many-worlds interpretation, holds that, when an observation is made, reality branches into multiple worlds, one for every possible result in the measurement.[2] As a result, it is not only the wave function that is smeared across possible states but also all existence. You exist through infinite multiverses, a reverberating echo that is perpetually modified to explore all possibilities, each version of yourself entirely unaware of the infinite other versions.

Such theories attempt to salvage our intuitions about the realism of the world around us. *Realism* refers to the idea that the world around us is truly separate from us and exists independently of our engagement with it. Quantum mechanics violates the realist expectation that things exist with objective properties before they are observed. In 2022, three theoretical physicists, Anton Zeilinger, Alain Aspect, and John Clauser, were awarded the Nobel Prize for their findings showing that quanta lack definite spin properties before they are observed, much like the vase that doesn't have a weight until it is put on the scale.[3] These findings suggest that our reality is not locally real; it does not have a solid, stable, pregiven existence in specific locations before the act of observation occurs.

IT'S ALL RELATIVE

One moment I am staring at the palms of my hands, held an inch away from my nose. The next I am looking into the beaming face of my baby son in a state of unstoppable giggling. I repeat the process again and again, the hilarity of my sudden appearance seeming to never get old. According to our current scientific understanding, children of his age have not yet developed object permanence, the cognitive ability to know that objects and people continue to exist when they are not being directly observed, giving this game of peekaboo its laughter-inducing element of surprise. This is a way of experiencing in which only what appears to us is felt to exist. We outgrow this way of seeing the world as we come to learn that large-scale objects do in fact continue to exist when we do not observe them. When it comes to the fabric of reality itself, however, this childlike perspective may be more accurate than our adult intuitions of unobserved permanence.

In order to make sense of the behavior of the quantum world, we must forget the idea of solid particles moving through space. In its place, we can adopt the process-relational view I explained earlier. Imagine that the mystics are right and that reality is a single, unitive, undivided whole. Now imagine that that whole is made up of a vast web of pure interactions, of the whole interacting with itself, with no separate entities like particles existing to do the interacting. What we call a particle is an event, not a thing—a moment of interaction. This is a reality without solid substance. It is a process of insubstantial occurrences shimmering in and out of existence to weave together the fabric of reality. The parallels between the insights of Eastern mystics and Western science have been powerfully explored in Fritjof Capra's *The Tao of Physics*.[4]

In recent years, this kind of perspective has been taken up by the physicist Carlo Rovelli, a highly respected mainstream scientist and popularizer of physics. He articulated it in his book *Helgoland*, where he calls it the "relational interpretation" of quantum mechanics.[5] He argues that all that exist are relations, which can be thought of as interactions, connections, events, or occasions but with no solid entities with their own independent existence behind these interactions. Reality is like the reflections in a hall of mirrors, where no solid mirrors actually exist.

In the relational view, the observations made in the laboratory that influence the properties of quantum systems should be thought of as interactions, or ways of relating to the system in question. Importantly, this does not require consciousness. Any physical system interacting with another physical system can establish a relationship or an "observation," although the latter term feels less applicable here—what we call an *observation* in a scientific experiment is just one example of a relationship being established between one physical system (the scientist) and another (the physical processes being studied). The moon continues to exist when not observed by a human, for example, due to its being part of the web of relations that make up reality.

We can also describe this view of the quantum world in terms of information. The scientific concept of information was developed to quantify signal transmission over phone lines. Information only makes sense when we are considering relationships, such as those between transmitter and receiver, making it well suited to this relational view of the world. It does not make sense to speak of information as existing at a single location without reference to anything else. Rather, informational dynamics play out in the structure of a whole system. The theoretical physicist John Wheeler coined the term *it from bit* to capture the idea that existence is fundamentally informational. Wheeler is also known for making the point that we live in a "participatory universe."[6] We are participators in existence; we are not fundamentally separate from the world around us.

Wheeler reached a radical conclusion with this perspective, arguing that it is our existence as conscious observers that collapses the possible states of the universe into actual ones, bringing our universe into existence. You created the big bang in this view by observing the world around you. Prior to the existence of conscious observation, the big bang and everything that followed it existed only theoretically, as potential events in a probability cloud. While I do not agree with this part of Wheeler's interpretation, it points to a fascinating aspect of the quantum world. Due to the indeterminate nature of quantum systems and the nonexistence of their states before they occur in the form of relations, we participate in the continual coauthoring of reality through our interactions with the wider world. An analogy has been made to the game twenty questions, where one thinks of an object and people ask yes-or-no questions to determine what the object is. Except in the case of the universe, there is no object; there is just the list of yes or no answers, and that description defines what the universe is. We ask if a particle is spinning up or down, and we get an answer that was previously undetermined. We ask another question of reality, and we get another answer. Reality emerges as the self-consistent structure that arises from the answers to these questions or interactions.

ENTANGLEMENT AND HOLISM

Quantum theory thus reveals a basic oneness of the universe. It shows that we cannot decompose the world into independently existing smallest units. As we penetrate into matter, nature does not show us any isolated "building blocks," but rather appears as a complicated web of relations between the various parts of the whole.—FRITJOF CAPRA[7]

How does a consistent world emerge from the relations of the quantum mechanical world? The answer may be found in the phenomenon of entanglement. Entanglement arises when two or more particles interact in such a way that their properties become correlated. When this happens, the properties of one particle are no longer independent of the properties of the other particle, even if they are separated by large distances. This phenomenon has been referred to as "spooky action at a distance," and it has been demonstrated experimentally.

In order to preserve our intuitions about the fundamental nature of space-time, scientists speculated about possible causal mechanisms whereby a change in one particle could instantaneously affect the state of the other particle. One idea was that a signal could be transmitted across space to influence the other particle, while another was that the states are determined before the particles are separated. These were both disproved in 1964 by a scientist by the name of John Bell.[8] Given the separation of the particles in some entanglement experiments, the hypothetical signal would need to travel faster than the speed of light to influence the second particle, violating Einstein's universal speed limit. What's more, entanglement cannot be used to transmit information as should be possible if a signal was being transmitted. Something other than signal transmission or causal influence seems to be going on here.

Immanuel Kant argued that space and time are not fundamental but are instead constructs in our experience that we use to navigate the unknowable real world.[9] If we take this idea seriously, then it makes sense of the way that entangled particles can show structured patterns of behavior across what appear to us as vast spatial distances. Given that the quantum mechanical world is more fundamental than the classical physical world, it may be that quanta do not actually behave in space, and instead of seeing their true behavior clearly, we force it through the distorting structure of our conscious experience, with its assumptions of the objective reality of space and time.

If this is the case, then there still would be something in reality beyond our experience of it that corresponds to the three dimensions of space, but it just wouldn't be a fundamental and absolute feature of our reality. We can imagine that, when quantum processes aggregate to form life-forms like us, the aggregated system that we call the organism can move independently in four different ways. This is what a dimension is, a way in which something can change that is independent of other ways in which it can change. I can move my hand up and down without moving it left and right, making these two axes independent, or orthogonal. They are two different dimensions. Without needing to know the true number of dimensions in existence, we can say that four exist with respect to the behavior of living systems. The one we call time is an inescapable process of change that the organism is continually subject to. The other three, those of space, are three dimensions of behavior, three independent ways in which the organism can change itself in its search for energy to sustain itself. These dimensions are relevant at our scale, but a different number may exist at the fundamental level.

If space is not fundamental, then there is no issue with entanglement. We might ask what the mechanism is that creates this pattern of correlations between the two particles. We currently know of no mechanism that could account for this behavior. My intuition is that, if science can find an answer, then it will be in the behavior of the whole system of reality rather than a local mechanism.

From the relational perspective, entanglement creates a single reality by connecting the properties of different particles. In this view, however, these properties are not preexisting features of independent objects. Rather, the properties of entangled particles are relational, meaning that they are defined by their interactions with other systems. Entanglement is a three-player game. When a scientist observes the correlation in states between the two entangled particles, it is only at that moment that the web of relations is formed and the pattern moves from the realm of possibility to the realm of the actual. The joint properties are only correlated regarding the third entity, the observer, who interacts with them both, weaving a web of interactions or relations.

We can also interpret this using the concept of information. Information is correlation. We say one thing carries information about something else if there is a relationship between the two that allows us to gain knowledge about one thing by observing the other. If there is no correlation between the two, then no information can be extracted about one object by studying the other. This implies the existence of the third participant, the entity who does the observing and establishes the informational relationship between the two objects. From the perspective in which only information exists, the pattern between the particles consists of a chain of informative interactions that include the observing entity. It is this chain of interactions that underlies the connection, not any causality in space and time in which one particle directly causes a change in the other.

Entanglement seems to be what gives us a coherent single reality that we all share. It weaves these interactions into the seamless whole that we call the universe. Consider all the particles that you have ever come into contact with in your life. Now consider the fact that, wherever they have ended up in the universe, they may still be entangled with parts of you. This is one way in which you can see your deep interconnectivity with the rest of existence. Entanglement reveals the holistic nature of reality, in which the properties of a system as a whole cannot be understood by analyzing the properties of its individual parts. This is because the properties of entangled particles are defined in relation to each other, and they cannot be separated or understood independently.

The approach of reductionism, of breaking things down and analyzing the component parts, has been powerful in building an understanding of classical physics and in building technology, but the scientific endeavor to understand quantum mechanics has revealed its limits. With reductionism, we believe we are unweaving existence to the point where all that will be left is individual separate threads, only to discover that, when we look at the world with maximal precision, we are confronted not with isolated elements but instead with the woven pattern of the whole. We are left with a vision of radical inclusivity. Our reality does not consist of separate things with their own causal power to influence other things. Existence is a single web of relations that evolve together as a whole; a dynamic network, forever emerging and becoming; an eternally blooming flower suspended in the void. The only "atom," the only true indivisible, is the entire universe.

We've seen that the findings of experimental physics begin to make sense when we adopt the view of nondual naturalism, in which separation and solid substance are seen to be mental constructs with no objective reality. Instead, our universe is understood to be a single relational process. I have now laid a foundation that can be used to understand consciousness, but at this point, we are no clearer on where it fits into our picture of existence. These insights into the nature of reality initially arose through the study of direct experience, before being validated by science. Is this a clue that there could be some fundamental connection between consciousness and wider reality?

SIX

IS CONSCIOUSNESS FUNDAMENTAL?

I regard consciousness as fundamental. I regard matter as derivative from consciousness. We cannot get behind consciousness. Everything that we talk about, everything that we regard as existing, postulates consciousness.—MAX PLANCK[1]

THE SOLITARY SKEPTIC

Every night, our minds conjure fantastical worlds in our dreams. Typically, these worlds feel convincingly real, and we rarely question their legitimacy when in the dream state, even when the narrative is totally illogical or absurd. If the avatar of yourself in the dream wanted to investigate the nature of the world around it, how would it go about doing so? Say you dream you are in a house. In this scenario you may inspect the walls and have the experience of them being made of bricks or perhaps even marshmallows because this is a dream. Say you do have the brick experience, however, and the dream version of you is reassured that the wall and the bricks actually exist. Then you wake up and realize that no real wall existed, nor did the bricks. Despite your seeming experience of inspecting a physical environment around you, everything in the dream was made of consciousness. Each morning we wake up from a dream world made of consciousness into waking reality. Is it possible that our waking reality could also be made of consciousness, just like the dream?

As I explain in the first chapter, we never directly experience matter, but the existence of mind appears undoubtable. Even if you doubt its existence, this experience of doubt is an occurrence in the mind, proving its existence. If we want to know what exists and we are strictly empirical about it, in the sense of only relying on what can be directly observed, then we are left with the conclusion that one's own mind is the only thing that can be known to exist for certain. Everything you've ever experienced could be a mere appearance in the only mind in existence: yours. This belief is known as solipsism.

When it comes to gathering an understanding of the world, we are in a sense imprisoned by our conscious experience. If we aspire to be strict empiricists, then we are stuck with solipsism and must give up on the project of trying to understand what is going on more broadly, as nothing beyond your experience could ever be known directly. Few people settle on a solipsism of this kind. It is more common to accept that a world might exist outside your personal experience, as it seems to, and that trying to infer the nature of that world is a reasonable project, even though we never come into direct contact with it.

What is the nature of the world that we think exists beyond our conscious experience? In the previous chapter I show how, rather than being made of a solid material substance that is independent of us, the material world appears to be a web of interactions in which we are embedded. When we interact with the world around us, this interaction often appears to us as a conscious experience. Some claim that these interactions are actually conscious experiences, arising in a reality in which only consciousness exists.

A MIND-ONLY REALITY

I explained in the opening chapter that idealism is the stance that only conscious experiences, or mental ideas, exist. I also described how, in the West, this position is associated with the eighteenth-century thinker Bishop George Berkeley.[2] Berkeley denied that anything actually exists in the world beyond our conscious experience of it. When you see a chair, there is the mental image of the chair, but there isn't actually a corresponding physical thing out there in the world. According to Berkeley, the whole of existence is made of mental images, a vast hall of mirrors with nothing behind the reflected appearances.

What would Berkeley make of the question "If a tree falls in the woods and no one is around to hear it, does it make a sound?" The way in which this question is answered hinges on the definition of the word *sound*. Typically, *sound* refers to the conscious experience of hearing something, so we can use this definition for the term *sound* and save the phrase *vibrations in the air* to refer to the physical phenomenon associated with an event like a tree falling in the woods. If the tree falls, it will produce the vibrations in the air, but without a conscious being with the right kinds of ears, there will not the experience of a sound. To Berkeley, only the perceptual event of hearing the sound has any reality.

What happens to the gap between matter and mind from this perspective? If consciousness is the substance out of which the universe is made, then the experience of existing in a world of matter can be understood as just that: an experience and nothing more. In this view there is no substance called matter; what we call matter is just an experience in the mind. A problem for this perspective is explaining why our experiences of the world around us are consistent and reliable. If I put some flowers in a vase and leave the room, I wouldn't be surprised if they were still there when I returned later. We typically have the intuition that this is because the flowers and the vase actually exist in the world, beyond our personal experience of them. To Berkeley, what exists in the world outside of our own minds are the minds of others, and this includes the mind of God. This is how he gets around the issue of the flowers continuing to exist when no one is looking at them, because God is always watching. Berkeley himself called this stance immaterialism to emphasize the lack of a solid substance behind our experiences, but it has come to be known as subjective idealism.

THE PROBLEM WITH IDEALISM

Berkeley's subjective idealism can be contrasted with objective idealism. While Berkeley denied there was an objective world beyond our personal experiences, objective idealists believe there is a world outside of our perceptions but that it is made of consciousness, not matter. To the objective idealist, the vibrations in the air that are produced when the tree falls do exist, but they are made of consciousness. Similarly, behind the appearance of the flowers is a consciousness that they are made of, a consciousness that continues to exist when you leave the room. In this approach only a single universal consciousness exists, but it somehow becomes fragmented into different minds and objects. According to this view, matter is merely how one fragment of universal consciousness appears to another fragment. Today, analytical idealism, a popular version of this kind of position, is associated with a thinker by the name of Bernardo Kastrup.[3] Kastrup suggests that a psychological process of dissociation, as occurs in multiple personality disorder, leads to the single universal consciousness breaking up into different alter egos, or alters. You and I are two of these alters of cosmic consciousness, according to this view.

What does it mean to suggest that consciousness is the substance out of which existence is made? We have seen that, historically, substance-based explanations have been placeholders for genuine insight into the nature of a given phenomenon. What's more, substance-based explanations leave us without an understanding of why the phenomenon in question exists; in the idealist perspective, consciousness is simply a brute fact. This is not a satisfactory solution to the mystery of consciousness. A major issue with claiming that reality is made of any substance, whether it be consciousness or matter, is that positing an uncaused substance that just exists for no reason is supernatural thinking and doesn't give us insight into what reality really is. You may as well say God made the universe in seven days if you are inclined to say the universe is made of some substance without explaining how or why.

If one digs into these substance claims, however, they usually turn out to be claims about emergence, about whether matter or mind is more fundamental than the other, with the less fundamental one emerging from the more fundamental one. If I claim that reality is made of the same thing that my individual mind is made of without calling that thing a substance, then what I am saying is that whatever the mind is occurs prior to the physical world. This issue of the relationship between the physical and the mental is what we must grapple with.

A challenge for these idealist perspectives is explaining how conscious experience relates to brain activity. If everything is made of consciousness and matter is an appearance with no causal power, then why does the operation of the brain seem to influence the contents of consciousness? An appearance shouldn't possess power of this kind. This would be like the image in the mirror determining how you look rather than the other way around. The idealist perspective would be plausible if the physical organization of the brain had no influence on experience, but this is not the case.

The neuroscience of perception is now very advanced, and we can predict aspects of the conscious experience of an individual based on the state of the brain. For example, in one study, researchers were able to reconstruct aspects of a visual movie that the participant was watching based on observing their brain activity—a form of technological mind reading.[4] What's more, we can induce experiences by stimulating the brain. It is very clear that the activity of the nervous system affects the contents of human consciousness.

The objective idealist has two options for how to respond to this issue. One option is to say that the seemingly physical brain is just the appearance of the conscious experience from the outside—it is just what conscious experience looks like to another conscious observer. This approach ignores the evidence that the state of what we call the physical brain influences the contents of consciousness in a reliable and predictable way, and it does not address why this relationship exists. I explain later that we can think both about how the brain affects consciousness and how consciousness affects the brain, but claiming that the interaction only goes in one direction does a disservice to our understanding of the fascinating and intricate dance between the two.

The other option for the objective idealist is to acknowledge the evidence of this relationship and to say that what we call the brain can influence consciousness because it is *made* of consciousness. If this option is taken, then we are back where we started, with an explanatory gap between what we call the physical brain (even if it is made of consciousness) and experience. If brain cells, neurotransmitters, and electrical activity are all made of consciousness, then we still lack an explanation of why their organization influences the contents of experience. It is this relationship between physical organization and experience that we must focus on in order to close the mind-matter gap.

PANPSYCHISM

All that physics gives us is certain equations giving abstract properties of their changes. But as to what it is that changes, and what it changes from and to—as to this, physics is silent.—BERTRAND RUSSELL[5]

One moment you feel invigorated by an intense charge; the next you disperse into a fuzzy wave of possible ways of being, smeared across the very substrate of existence itself. You collapse into a point again and feel yourself being dramatically pulled around by other things like you. There is no time here. You exist in an eternal buzz of being—no hopes, no pains, no boredom, just this jarring ordered chaos of existence. Is it possible that it feels something like this to be an electron?

In the reductionist framework, nothing genuinely novel can ever come into existence. If reductionism is correct, then either conscious experience must be an illusion or, if it genuinely exists, it must also be present at the most fundamental level of existence. This logic was the inspiration for a position related to idealism called panpsychism, which holds everything in existence to be conscious, even subatomic particles. In this view, matter and mind are equally fundamental; both exist as two sides of the same coin. Everything in existence has an external material aspect and an internal conscious aspect according to panpsychism. After posing the hard problem of consciousness, the philosopher David Chalmers took up the panpsychist position, arguing that there was no way that consciousness could emerge from nonconscious matter. He posited the existence of consciousness as a brute fact, being present in everything in existence.

Some feel that, if every particle of matter is conscious—a version of panpsychism known as micropsychism—then that explains how it is possible for you to be conscious. You may have noticed, however, that you are not a particle. The way that many conscious particles could be able to aggregate into the larger consciousness of a whole organism remains, and is known as the combination problem. This approach is ultimately unsatisfactory. We are left without an explanation of why consciousness itself exists in the first place, as well as why your consciousness in particular exists.

In recent years, the philosopher Philip Goff has championed panpsychism, particularly in his 2019 book *Galileo's Error*, the title referring to Galileo's attempt to confine science to objective reality by declaring that the "Book of Nature is written in the language of mathematics."[6] According to Goff, this was a mistake, and panpsychism offers a way to bring consciousness back into our scientific picture of the world. In this framework, reality itself *is* consciousness.[7] Everything in existence, like you and I, is inherently qualitative from the inside but has quantitative properties when looked at from the outside, as our bodies do. Goff points to the work of the philosopher and mathematician Bertrand Russell, an intellectual giant of the twentieth century, when describing this version of panpsychism. According to

Russell's view, known as Russellian monism, we shouldn't think of particles as being conscious but as being consciousness. Rather than having the capacity to be aware, consciousness is the very substance that constitutes them.

In this framework, reality itself is consciousness, much like with objective idealism. The main difference between this position and objective idealism is that panpsychism often focuses on the parts that existence is made of, the particles, being made of consciousness, while objective idealism often starts with the whole, a cosmic universal consciousness. Another difference is that, under panpsychism, matter and mind are equally fundamental, while the idealist holds the appearance of matter to be derivative of mind. The consciousness of particles is often described as protoconsciousness, a minimal version of experience that may not feel like much at all. This makes this stance more palatable to many who feel uncomfortable with the idea of our rich conscious experience being possessed by something as simple as a particle of matter.

One issue with this approach, however, is the nature of particles of matter. As I explored in the previous chapter, twentieth-century physics has disabused us of the notion that particles are tiny, solid building blocks that have their own permanent existence when not interacted with. As a result, it is hard to say there are individual things present at the subatomic level that could be conscious or made of consciousness. While this is a challenge for micropsychism, this vision of the physical world actually fits nicely with idealism, as the world of subatomic particles does seem more like Berkeley's immaterial vision of a world where only appearances rather than solid material substance exist. This correspondence does nothing to resolve the broader problems, described earlier, that idealism faces, however. Seeing what we call particles as relational interactions does take us to an interesting related position, however, called panexperientialism.

PANEXPERIENTIALISM AND PROCESS PHILOSOPHY

The misconception which has haunted philosophic literature throughout the centuries is the notion of "independent existence." There is no such mode of existence; every entity is to be understood in terms of the way it is interwoven with the rest of the universe.—ALFRED NORTH WHITEHEAD[8]

While a student at Trinity College, Cambridge, in the 1890s, Bertrand Russell had a tutor by the name of Alfred North Whitehead. The two would later collaborate on *Principia Mathematica* in the first decade of the twentieth century, a hugely important work that put mathematics on a firm footing based in philosophical logic. By the 1920s, physics was undergoing a revolution with the development of quantum mechanics. In the same year, in the midst of this revolution in our understanding of the fabric of reality, Whitehead traveled from Harvard to Edinburgh to give the Gifford lectures. He proposed a new metaphysics that attempted to make sense of the nature of reality and would shed light on the revelations coming from quantum mechanics.

Decades later, Ludwig Wittgenstein, who was Russell's student, would write, "[I]f a lion could speak, we could not understand him."[9] Something similar seems to have happened at these lectures but with a philosophical giant speaking instead of a lion. Apparently 99 percent of the audience chose to not return for the next lecture after attending the first, with the audience size collapsing from six hundred attendees to just six for lecture 2. Fear not, however: I only give the broadest possible overview of his worldview here.

At the core of Whitehead's philosophy is the belief that reality does not consist of a solid substance but should instead be conceived of as an unfolding process of becoming.[10] This becoming occurs between two poles: one of potentiality and one of actuality. The future is an open, undetermined field of possibility in this view, and in each moment, particular events arise out of that space of possibility, in a manner that depends on events that have occurred in the past. The events that actually occur, termed *actual occasions* by Whitehead, are what everything around us is made of. Reality is a process of ever-changing events, not something with a solid material foundation. When particles and their properties are observed, these, too, are seen as events or occasions rather than solid particles with their own existence independent of the observation.

In this framework, these actual occasions are occasions of experience. Everything that exists is seen as a form of experience in this perspective, making it an example of panexperientialism. The difference between the use of the word *psyche* and *experience*, as in panpsychism and panexperientialism, is subtle here, but we can think of experience as a simpler form of subjectivity than the psyche, which often connotes a fully formed mind, although this is not usually how it is intended when used in the term *panpsychism* these days. As with panpsychism and idealism, however, the issue here is that experience or consciousness is simply stated to exist without being accounted for.

It is easy to see why many have felt that the findings of quantum mechanics support metaphysical idealism. If all that exists are interactions and one example of those interactions is conscious observation, then perhaps conscious observation is the nature of every interaction. In making this move, we are generalizing from our experience to the nature of the physical world that lies outside our experience. We see here that it is the investigation of our own experience itself that commonly serves as the inspiration for adopting the stance of idealism. What is it about the mind that would lead someone to the seemingly radical conclusion that all of existence is made of consciousness?

FROM MYSTICISM TO IDEALISM

The Real is ever-present, like the screen on which the cinematographic pictures move. While the picture appears on it, the screen remains invisible. Stop the picture, and the screen will become clear. All thoughts and events are merely pictures moving on the screen of Pure Consciousness, which alone is real.—RAMANA MAHARSHI[11]

You sit in deep meditation, losing connection with your everyday experience of the world. Suddenly all separations fall away, and there is only a dance of appearances arising in awareness. The feeling of the body dissolves into a continuous cloud of sensations made of vibrant, ephemeral sparks of being. Nothing feels substantial; the only ground here is the pristine, untouchable clarity of awareness itself, which shines like a transparent, immovable mountain of light. Nothing you have ever experienced in life has felt as real as this. All else is impermanent change. However, this not only feels like the foundation of what you are but also like the foundation of all of existence.

Idealism and panpsychism are very popular in contemporary spiritual circles, but why should this be the case? Let us consider the change in perspective that can occur with the mystical experience. We could describe our usual commonsense picture of the world as a combination of naive realism, naive materialism, and epiphenomenalism. Naive realism is the belief that the way we perceive the world, as consisting of separate objects, reflects the way the world actually is. Naive materialism is the idea that our experience of solid substance also reflects the way the world actually is. Epiphenomenalism is the idea that mental experiences are less real than physical matter. Mystical experiences do a good job of bringing all three beliefs into question.

The mystical state invites us to question naive realism by showing our experience of the world around us to be just that: an experience rather than a direct perception of the world as it is. It also dispenses with naive materialism by showing solid substance to also be an experience rather than a feature of the world. Finally, in the mystical state, the contents of consciousness are typically felt to be more directly real than the abstractions of physics. This does a good job of dispelling the idea that consciousness could be an unreal epiphenomenon. What's more, during such states conscious experience is often felt to be deeply grounded, possessing an unchanging core of formless awareness that feels more stable than anything else in existence, something that we would not expect of an irrelevant by-product of physics.

In the mystical state, separations are seen to be mental constructs, and as a result there is the sense that one is the same thing as the rest of reality. If you take consciousness to be the stable core of what you are after such an experience and combine it with the intuition that you are the same thing as the rest of the universe, then you may come to the conclusion that consciousness is what the universe is made of. This logic, of the individual conscious mind being equivalent with a universal consciousness that forms the ground of existence, is found in the Vedic texts of Hinduism. There is reason to believe that the authors of the Vedas took nonordinary states of consciousness as their inspiration for this conclusion. The opening lines of these texts inform us that those who wrote them routinely consumed a ritual drink called soma, which they describe as having psychedelic properties: "We have drunk the soma; we have become immortal; we have gone to the light; we have found the gods. What can hostility do to us now, and what the malice of a mortal, O immortal one?"[12] We do not know for certain what soma was, but according to these texts, its mind-altering effects produced many of the philosophical ideas documented within.

A few hundred years later, Plato, the father of Western philosophy, would consume a sacrament of this kind and would come to similar conclusions. Plato took part in the Eleusinian mysteries, ancient Greek religious initiation rites in which a reportedly psychoactive potion called the *kykeon* was consumed. He speaks of the mysteries in his work *Phaedo*, in which he also explores the idea of the immortality of the soul. For Plato, true reality consisted of a nonmaterial realm of perfect forms or ideas, with the material world being a corrupted reflection of this sphere of spiritual perfection. According to Plato, our minds participate in this realm of forms, but our bodies are disposable additions, allowing consciousness to survive the death of the flesh.

In both ancient Greece and Vedic India, these idealist visions of a reality in which consciousness survives the death of the body are both described as being inspired by the experiences brought on by psychoactive potions. We do not know what the soma plant was, although some have speculated that it may have been a species of psychedelic mushroom. More is known about the ancient Greek potion, however. As documented in *The Immortality Key*, recent archeo-chemical evidence from a *kykeon* cup found in a Greek colony in modern-day Spain found traces of ergot, a fungus from which LSD is derived.[13] This raises the possibility that Plato's potion may have been truly psychedelic, potentially producing the kinds of mystical experiences people are having today in research trials.

In 2021, researchers sought to test whether psychedelic-induced mystical experiences really do produce changes in metaphysical beliefs.[14] They found that, post-psychedelic experience, participants were more likely to believe that consciousness is a fundamental feature of existence rather than arising secondary to matter. This research seems to show that these mystical experiences really do have the power to shape our beliefs about the nature of reality, moving them in the direction of believing consciousness to be more fundamental than we typically think.

IS AWARENESS FUNDAMENTAL?

Consider the fact that no matter how many planets and stars are reflected in a lake, these reflections are encompassed within the water itself; that no matter how many universes there are, they are encompassed within a

single space; and that no matter how vast and how numerous the sensory
appearances of samsara and nirvana may be, they are encompassed within
the single nature of mind.—DUDJOM LINGPA[15]

As you look at this book, ask yourself, "What is it made of?" As a matter
of experience, the book that you perceive is actually made of consciousness.
While there is a physical thing out there in the world that corresponds to
this image of the book, it does not possess the properties that the image
of the book does. This may seem surprising, but we can see why this is the
case if we consider the appearance of the sun. To us, it looks a certain way
as a result of the structure of our eyes and brains, but it would look very
different to a creature with eyes that were sensitive to other wavelengths of
radiation. Which appearance is correct? What is the true appearance of the
sun? There is no true appearance; there is an energetic process that looks
like nothing in itself but has a certain appearance when viewed by a conscious system. The world *as you experience it* is indeed made of consciousness, yet there is also a world beyond your conscious experience.

This position is known as transcendental idealism and is associated with
the philosopher Immanuel Kant. While the label *idealism* may suggest a
belief that only consciousness exists, this is not the case in Kant's vision of
reality. This perspective is considered a form of idealism because it holds
that the world we experience is dependent on the structure of our minds,
but in transcendental idealism, there is also a world that exists beyond our
experience of it. Kant called the world of our experience the "phenomenal"
world and the world in itself that exists beyond our minds the "noumenal"
world. The term *phenomenon* comes from the Greek for *thing appearing to
view*, while *noumenon* comes from *something conceived*. Kant used the term
noumenon because he believed we can never know the thing in itself directly
and must instead infer its existence; we must conceive of it.

This insight from Kant is crucial in understanding the nature of consciousness. We may think of consciousness as the way we come to know
the world around us, but from the Kantian perspective, this isn't actually
the case. When we experience something, such as the blueness of the sky,
we are not coming to know the true nature of the sky; we simply believe
a blue sky to exist. We can think of the contents of consciousness as beliefs in the character of the world, a world we believe as consisting of

objects with qualities that do not actually exist in the way that we perceive them. A conscious system is one that is capable of believing in a world of qualities.

For Kant, we are disconnected from the real world in its true nature. Instead, all our seeming contact with reality is mediated through phenomenal experience in consciousness. A later philosopher, Arthur Schopenhauer, argued that it might somehow be possible to access noumenal reality through our subjective experience, as we are ourselves an instance of the thing in itself. From this perspective, it would make sense if the consciousness that makes up your experience were also the nature of reality in itself. If that were the case, then it would explain how one could access true reality through one's own direct experience.

Earlier, I broke consciousness down into the conscious contents and the formless awareness in which they arise. The accounts of idealism and panpsychism explored here take the qualitative contents of consciousness to be fundamental, and I explained why this does not successfully close the mind-matter gap. What about the formless awareness that is felt to not be bound by concepts or time? Might that be fundamental? After all, it does appear to be unaffected by profound changes to the brain, such as high doses of a psychedelic. Could formless awareness be our portal to the noumenon, to reality in itself?

The formlessness of awareness is not only an intellectual concept to believe in or not; it is a description of a facet of experience that appears to many to lie at the core of consciousness. Right now, you can try to notice this formlessness. Try repeating a random word to yourself, either in your head or out loud. Notice that you are aware of the sound of the word. Now bring your attention to the awareness in which the sound is arising, the experiential space in which it is heard. Strangely, this space seems to be there even when there is silence. What's more, it is the same space in which all experience occurs and is completely unaffected by what arises in it. Nothing may have happened for you during this brief investigation of experience, but many who have devoted themselves to the internal study of consciousness have reported that they find this same formless awareness.

This formless awareness can be quite shocking when discovered for the first time. It appears to be featureless in itself, and it is this emptiness or absence that allows it to contain everything that arises in experience. Because it exists prior to our constructed mental concepts, such as subject, object, and even time, it appears to be still, unchanging, not bound by time, and nonfinite. It feels unimaginably grounded and stable, often feeling like the most real thing one has ever experienced. Coupled with this groundedness is the intuition of being fully part of reality that arises with the dismantling of conceptual divisions. Taken together, these intuitions of the groundedness of formless awareness and of being fully part of an undivided reality can give the impression that this formless awareness is what lies at the core of existence. This gives us a form of idealism where pure formless awareness is what reality is made of, rather than the qualitative forms that make up the contents of consciousness.

I personally do not think this version of idealism is justified. What are we talking about when we use the term *awareness*? Typically, when we are aware, we are aware of something. Even if we are not currently aware of anything, *awareness* is the term we use for the space in which experiences *can* arise. *Awareness* therefore appears to refer to a kind of dynamic or relationship in which experiential content can occur. We have no reason to believe that this capacity is fundamental and every reason to think it arises with creatures that can attempt to know the world around them, like us. To claim that awareness is fundamental without saying why is to make a supernatural claim. If it is fundamental, then no explanation could be proposed regarding why it is aware; it would just be a brute fact, like positing the existence of a creator God.

While the core of consciousness appears to be formless awareness, the core of formless awareness appears to be formlessness itself. It seems possible that it is formlessness, not formless awareness, that gives the latter its groundedness. In the metaphysical picture sketched earlier, formlessness is the fundamental ground of existence. It may just be that the same potent void, the interplay of nonbeing and being, of formlessness and form, underlies both matter and mind. In the case of matter, the forms are the structures of objective reality that physicists study. In the case of mind, the forms are the contents of consciousness. Both may ultimately be grounded in the same formlessness, however, accounting for idealist intuitions while keeping the physicalist insight that consciousness itself is emergent. By taking the formlessness that is at the core of consciousness to be fundamental, we can accept a very lean version of idealism.

A SYNTHESIS

The words we use to describe reality are just that: words. Each philosophical system I have explained can be thought of as a map overlaying reality, rather than a true description of reality itself. With this in mind, we can consider whether idealism, panpsychism, and physicalism might all be capturing valuable insights about the relationship between matter and mind without any of them solely being true. This creates an opportunity for a synthesis of these metaphysical positions.

Earlier, I put in place a picture of reality in which forms arise as a relational process out of a formless ground. I then showed that this worldview explains the behavior of the physical world. If we think of consciousness, too, as a relational process, then we can conceive of both mind and matter as different modes of operation of the same metaphysical substrate, as Spinoza suggested. What would this mean for the nature of consciousness? If consciousness is a relational process, then it must exist through the physical operation of the conscious system in question rather than being conceived of as an additional nonmaterial phenomenon produced by the brain. It would mean that it would be synonymous with the functioning of a whole organism, in our case, and not just brain activity. In this picture there is no fundamental metaphysical difference between mind and matter.

If consciousness is a relational process that arises out of a formless ground and so is matter, then the insight of idealists that reality is made of the same thing as consciousness is correct. It is just more accurate and naturalistic to say that both are relational processes arising out of a formless ground, as opposed to saying that matter is consciousness. What's more, the insight that the formless core of awareness is the same thing as the ground of existence is valid—the formlessness we find in experience is the same formlessness out of which the physical world arises. We experience this formlessness as the space of formless awareness, but if we were to say that at the bottom of reality is formless awareness, then we would be in danger of reifying this nonconceptual ground into an object: that is, awareness. Instead, we must sit with the unobjectifiable negative definition of it being formless. We could perhaps call it protoawareness, in the tradition of panpsychism, to emphasize its continuity with what we call awareness, as long as we are clear that it only underlies awareness rather than being aware itself. What it really is, however, is the nonfinite space of potential out of which everything arises. The idealist intuition that what one is at one's core is the same thing as the ground of existence is valid.

In addition to seeing these insights from idealism as useful in understanding the nature of awareness and panpsychism in understanding the continuity between matter and mind, this synthesis holds physicalism to be essential for understanding consciousness itself. The contents of consciousness clearly depend on the organization of the physical world. When you see an object, both the physical object itself and parts of your body, including your eyes and brain, determine what will be consciously perceived. This dependency is what is captured by physicalism. What is important in physicalism is not that the world is made of a solid material substance but the insight that the contents of consciousness depend on the patterns in reality studied by physicists.

Going forward, I can largely put aside the talk of formlessness. This concept is not necessary to understand the scientific account of consciousness offered here, but for the philosophically rigorous, it is good to have it operating in the background. Even with this idea of formless awareness being grounded in the very core of reality, I still have not explained where and why consciousness itself exists, with all its qualitative contents. This is the issue explored in the rest of this book.

SEVEN

BEING PART OF NATURE

From Animism to Human Exceptionalism

I'm truly sorry Man's dominion
Has broken Nature's social union,
An' justifies that ill opinion,
Which makes thee startle,
At me, thy poor, earth-born companion,
An' fellow-mortal!

—ROBERT BURNS, "TO A MOUSE, ON TURNING
HER UP IN HER NEST, WITH THE PLOUGH"[1]

EMERGENCE

If the space of qualitative experience we call consciousness is not fundamental, then it must have emerged at some point in the evolution of the universe. We've seen that at the bottom of our reality is an insubstantial fabric of relations, a holistic tapestry of connections in which we are fully embedded. Science has shown us that out of this fabric emerges patterns of interaction that we call chemistry, and out of chemistry emerges the complex dynamics we call biology. Saying that matter is fundamental is often taken to mean that it is what is most real. This is a tenent of the philosophy of reductionism, which arose with the related method of reductionism developed by Descartes. According to this view, the reactions of chemistry, the goals of biological systems, and your hopes and dreams are all "epiphenomenal." They are irrelevant to the basic functioning of the universe that all happens at the level of matter. Everything else is like steam rising from the engine of a steam train. They are unimportant sideshows, occurrences that have no real significance to the operation of the system in question.

There is another way of thinking about the fundamental nature of matter. In the nonreductionist view, you and I or a slug or an oxygen molecule are no less real than a subatomic particle, but the particles are more fundamental in that we depend on them for our existence. From this perspective, the complex phenomena that emerge within the dance of matter are very much real, but it would not be possible for the chemical compounds, the worms, the eagles, or your brain to exist without these particles. In contrast to the reductionist focus on parts alone, this view sees the reality of wholes, too, of the arrangements of parts into systems with their own existence.

In both the reductionist and the nonreductionist framings, it is possible to consider consciousness as emergent. To the reductionist, consciousness may be seen as a property of brains, but much like the brain itself, it is not seen as real in the same way that particles are real. To the nonreductionist, consciousness is as real as anything else in existence, but it may still be a property of brains or of other complex systems, such as artificially intelligent machines. Here, I take the nonreductionist view, seeing consciousness as no less real than matter.

Irrespective of the camp one falls into, the issue of emergence raises a tricky question: Which emergent systems are conscious? If we had an agreed-upon understanding of what dynamics in matter underpin consciousness, then we would have a foundation to determine which things are conscious and which are not. Without such a framework, there is no basis for us to make such claims about the capacity of others to experience. What's more, the subjective nature of consciousness means that it is impossible to directly observe and measure the capacity to experience of any system, and so we truly have no basis on which to make such claims. Interestingly, the common default stance on consciousness in the natural world isn't the rational conclusion of agnosticism regarding which systems are conscious, but instead of human exceptionalism and superiority over the rest of nature in the domain of experience. Conveniently, it is humans who proclaim that it is they who are uniquely special among all the forms in the universe when it comes to the ability to feel. How did this assumption come to be so pervasive, and is there any reason to believe it to be true?

HUMAN EXCEPTIONALISM

Organic life, we are told, has developed gradually from the protozoon to the philosopher; and this development, we are assured, is indubitably an advance. Unfortunately, it is the philosopher, not the protozoon, who gives us this assurance.—BERTRAND RUSSELL[2]

When considering where consciousness emerged, it would make sense to follow the story of emergent phenomena and consider at each stage whether this could be the point that experience came into existence. We begin with the fundamental particles of physics, out of which chemistry emerges. Inorganic chemistry is well described by the physics of particles that underpins it, giving the impression that there is no reason to think that consciousness came into existence at this point. In fact, I am unaware of this position—let's call it chemopsychism—ever having been proposed. At a higher level of complexity, organic chemistry emerges with life. We are living things, and we are conscious things, so it would make sense to consider whether this may have been the moment consciousness came into existence.

The naturalist Ernst Haeckel is perhaps best known for his beautiful scientific illustrations of natural forms. He also explored the question of which of these natural forms should be thought of as being conscious. He coined the term *biopsychism*, for the position that all living things are conscious, and *zoopsychism*, for the position that only animals with nervous systems are capable of experience. He himself rejected both ideas, deciding instead on *panpsychism*. In addition to these options, we can add the idea of radical human exceptionalism: the position that consciousness only came into existence with our species.

Much of contemporary thought about consciousness is informed by the assumption of human exceptionalism. For a long time, this was of the radical kind, in which we were seen as truly unique and different from the rest of nature in being the only conscious creatures in existence. Today, however, a less radical version dominates, in which we are seen as exceptionally conscious, and the idea of other creatures being conscious has been tolerated, as long as they are less conscious than we are.

If you are a pet owner, then you may find it surprising that scientists and philosophers still debate whether humans are the only beings capable of feeling. To the radical human exceptionalist, your dog or cat is just a robot. It may look like your dog is happy to see you, and you may think it is unethical to physically harm your cat because that would cause it pain, but in reality, they are as empty of feeling as a clock or a table, according to this perspective. As recently as 2012, a group of consciousness scientists felt it necessary to make a public declaration that pushed back against this idea. *The Cambridge Declaration on Consciousness* states that the "weight of evidence indicates that humans are not unique in possessing the neurological substrates that generate consciousness. Nonhuman animals, including all mammals and birds, and many other creatures, including octopuses, also possess these neurological substrates."[3]

From one perspective, human exceptionalism is reasonable when it comes to the study of consciousness, as you know for a fact that humans can be conscious by virtue of being one. You can then charitably grant that I am probably conscious, too, given our similarities as fellow *Homo sapiens*. It would be more surprising if I were not conscious, given our structural similarities (and the fact that I've written a book on consciousness), than if we were both conscious, so it is reasonable to assume the latter conclusion. A similar logic can then be applied to apes, monkeys, perhaps dogs, possibly whales and other charismatic megafauna, and so on. At this point, however, we have moved beyond human exceptionalism and are back in the place where agnosticism about the consciousness of such creatures would be the appropriate stance, until we have a sense of what it is that makes any given system conscious. By the time we get far from things like us, to octopuses, for example, we are left with failing intuitions. Without an accepted theory of consciousness, we can't say whether jellyfish or earthworms experience anything. Why, then, is the dominant stance one of human exceptionalism or at least human superiority? The idea that humans are superior to the rest of the living world when it comes to the capacity to feel is routinely presented as self-evident, but is this really the case?

ANIMISM

As I descended deeper into the limestone cave, the walls began to narrow around me. Passing a choke point deep below the surface of the earth, the walls suddenly opened into a cavernous space. Sealed for thousands of years until only a few decades earlier, this site had been chosen for use by the ancient people of what is now Portugal more than 50,000 years ago. I kept walking, and markings began to appear on the wall. In front of me were countless images of ancient horses, now-extinct aurochs, and other prehistoric fauna, some carved, some in charcoal, some in red pigment, all tens of thousands years old.

Paleolithic cave art like this is found around the world, and the images often depict wild animals. Sometimes, an explicit connection between the human and nonhuman world is depicted, as in the case of therianthropes. These are depictions of human-animal hybrids, a common trope in shamanic traditions. At the time of this writing, the world's oldest discovered cave art is from Sulawesi Indonesia, a minimum of 44,000 years old, and it depicts such human-animal hybrids. Several thousand years later, someone left a mammoth ivory carving of a lion-man hybrid in what is now the German Alps. The seeming universality of these themes has led many researchers to propose that the presumed sense of connection to the nonhuman living world these artworks are thought to depict may be an innate psychological stance of humans.

For most of human history, we lived as hunter-gatherers, our lives deeply embedded in the natural environments we inhabited. As David Abram documents in *The Spell of the Sensuous*, this way of living has a profound effect on our experience of the world around us.[4] People who live this way typically feel empathically connected to nature. Often, in such a perspective, humans are seen as one among many beings in the natural world, with no inherent superiority. The interrelatedness of all living and nonliving things is routinely emphasized, and as a result, so is the balance and harmony of the natural world. Cultures that relate to the more-than-human world around them in a respectful way are known as animist cultures.

In such a worldview, the sense of being a subject of experience exists widely. To an animist, a stag or an insect experiences the world around them much like we do. They are our kin and, as such, are worthy of respect. Animists may also treat specific nonliving natural phenomena, such as individual rivers and mountains, as if they were alive and capable of having their own perspective and concerns. This differentiates it from panpsychism, which holds that consciousness is a fundamental aspect of existence but doesn't commit to which emergent features of the world, such as waterfalls, stars, or beetles, become subjects in their own right.

THE MAKING OF THE MODERN WORLD

I write these words sitting in my home in a Europe that was very different a thousand years ago. Life was predominantly organized around what we now call feudalism, a system where the majority of people worked as serfs on land that belonged to a tiny minority of lords. In my native Britain, Christianity was pervasive among the landowning class, while animistic pagan beliefs were yet to be eradicated among the peasants.

Whereas today we arrange our economies around a single conception of value, as quantified by monetary worth, this was not the case in the pre-modern period. To the pagan populace, nature would presumably have not been seen simply as resources but as the enchanted realm of spirits, a place where fairies danced on toadstools and the spirit of the earth rustled the leaves. Our modern economies relate to nature as material that can be used for a variety of human-centered goals. If something has great usefulness as a tool for increasing our power to effect change in the world, then we say it is high in instrumental value; we can use it as an instrument to achieve our goals. Large rocks, for example, have high instrumental value if your aim is to cross a stream and you want to arrange some stepping stones to achieve this goal.

To a psychopath, the same thing could be said of humans who could be killed and arranged in the same way as the rocks. To the nonpsychopath, the human is not to be used as a means to an end in this way; they are seen as an end in themselves, as something possessing their own inherent value. A thousand years ago, it seems likely that nature would have been seen as possessing value in and of itself; it may have been perceived as kin in the same way as we perceive other humans. Somewhere in the transition from the medieval period to the modern period, this belief in the inherent value of nature appears to have been broadly lost in Europe, and Europeans shifted toward viewing these relatives as resources instead.

Many scholars point to the Black Death as the event that triggered the death knell for feudalism and the medieval period and set the stage for the emergence of the modern world.[5] The plague killed almost a third of Europe's population, a vast collective trauma that presumably left no individual untouched by its repercussions. The disruption this wrought on the feudal economy is thought to be vast, breaking traditional ways of living and pushing people to focus on their own basic survival needs. Peasants looked out for their own self-interests by demanding higher wages, which was possible due to the drastically reduced labor force. Instead of being tied to one particular plot of land, they became empowered to travel around in order to look for better opportunities for work, due to high need for labor. Upward social mobility started to occur for the serfs, to the point that the landowners introduced "sumptuary laws" to ban them from dressing like the nobility.

On the other side, the landowners looked out for their own needs by shifting away from labor-intensive grain farming and toward using the land as pasture for animals. They also took to forcibly enclosing plots that were previously communally owned to increase their landholdings and thereby their power. Before this process, serfs tended the land using the ancient open-field system, which prioritized sustainability over productivity. We see here a shift away from traditional sustainable practices and the holding of communal land to extractive practices focused primarily on productivity for the benefit of private individuals. When it comes to understanding the transformation in Europe's economic relationship with nature, this is arguably the key change we must consider. This shift in the material engagement with nature may have also served to change how the people of Europe came to see the rest of the living world, influencing the philosophers who would establish how we think about consciousness today.

Humans have a seemingly innate capacity for empathy. If we believe something to be like ourselves, then we will feel its pain when it suffers. The pagan worldview of a nurturing Mother Earth among the peasants was presumably a barrier to the emergence of extractive practices that aimed to plunder nature to maximize the power of landowning individuals. As Caroline Merchant puts it, "One does not readily slay a mother, dig into her entrails for gold, or mutilate her body."[6] To empathize with something is to presume it has the capacity to feel, to see it as a subject. When we prioritize ourselves and withdraw our empathy, we go from seeing others as subjects with their own experience and see them instead as objects. In order to make a new extractive relationship with nature possible, nature had to be turned from a subject to an object in the minds of Europeans.

THE BIRTH OF SCIENCE AND THE MIND-BODY SPLIT

By the seventeenth century, Europe was undergoing a great transformation. Vast wealth flowed into the continent, stemming from the plundering of nature and the enslavement and exploitation of other humans by the imperial and merchant classes. It is in this context that our modern idea of the natural world as an unconscious mechanistic machine, as a collection of unfeeling objects, rose to prominence. The father of the scientific method, Francis Bacon, was a deep believer in the superiority of humans over the rest of nature. Bacon described Mother Nature as a "common harlot" who should be "put on the rack" and have her secrets "tortured" out of her. Torture was something Bacon was familiar with in his role as attorney general, and it was a practice he employed against the people of his country.

Descartes would follow in the footsteps of Bacon, offering both a method by which nature could be interrogated and a worldview that made its plunder for profit more palatable. Both Bacon and Descartes were Christian philosophers who lived in places that believed in the Bible's paradigm of human exceptionalism. In keeping with this, Descartes argued that only human beings are conscious. According to Descartes, in all of nature, it is only we who feel pain and happiness, who know sensations like touch and taste; all other living things are unfeeling mechanisms. Humans, he believed, are conscious by virtue of their divine connection to God; we alone matter, and as a result, we are permitted to treat nature as we please. Descartes helped make the plunder of nature possible through this picture of living things as objects rather than subjects.

Descartes argued that, if we wanted to understand an object, we should ignore the characteristics of the whole and focus our attention only on the parts. He took to torturing animals, including his wife's dog, in order to examine the mechanisms that give the impression that they were in pain. If one looked at the whole animal, then one might see it exhibiting what appeared to be a state of agony. If one looked at the muscles, however, then one could see that at this level there was only mechanism. Descartes convinced himself that animals only gave the impression of experiencing such subjective states as pain. If we only look at the components alone, then there is no sign of the pain.

This view of nature as mechanical and unconscious is what gave rise to the modern incarnation of the mind-body problem. If the body is unconscious mechanism, then how is it that we do in fact experience such feelings as pain? Today science has rejected Descartes's explanation that humans are conscious by virtue of possessing a divine immaterial soul, but it has kept the assumption of human exceptionalism. It is not uncommon to hear scientists say that certain creatures can't be conscious because we can understand their functioning in a purely mechanistic manner, overlooking the fact that this is true in theory for humans, too, and yet, we are still conscious.

Descartes's ideas were accepted because they were useful in the conquest of power, not because they were correct. The flawed assumption of the unconsciousness of nature gained popularity as it justified the accumulation of vast amounts of wealth through this new economic system. This intellectual legacy has left modern science with unexamined assumptions and biases toward human exceptionalism that blind us to a true understanding of consciousness.

SEPARATING SCIENCE AND SPIRITUALITY

When the animate powers that surround us are suddenly construed as having less significance than ourselves, when the generative earth is abruptly defined as a determinate object devoid of its own sensations and feelings, then the sense of a wild and multiplicitous otherness (in relation to which human existence has always oriented itself) must migrate, either into a supersensory heaven beyond the natural world, or else into the human skull itself—the only allowable refuge, in this world, for what is ineffable and unfathomable.—DAVID ABRAM[7]

Throughout human history, spiritual experience and our understanding of the world were deeply intertwined. It was Descartes who definitively split these two apart in the West. In Descartes's worldview, the material world was lowly, devoid of value, to be exploited. The divine, however, was related to an otherworldly heaven. Science and religion were given the corresponding tasks of managing the resources of nature and of mediating with this transcendent divinity. While the change in economic relations with the earth following the Black Death can be seen as the proximate cause of our disidentifying with nature, this move toward a transcendent conception of divinity can be understood as a deeper cause, one that has been operating for thousands of years.

Our modern secular culture is an anomaly in human history. Most cultures have had some sense of the divine or the spiritual at their core. It is in terms of the divine that most cultures frame their understanding of the value, or lack thereof, of nature. Today, however, we typically hear these terms as indicating some additional, magical, supernatural property that is not to be found in our scientific understanding of the world. There is another way of thinking about the experience of the divine, however, that is entirely naturalistic and fits with science. I first unpack this understanding of the divine so I can interpret how different cultural traditions have seen nature and how we should relate to it.

I must return here to the claim that reality itself is nonconceptual: It exists prior to our ideas about it. "The map is not the territory," as philosopher Alfred Korzybski put it.[8] When we drop our concepts and experience this clearly, we also drop our sense of the world as being mundane and familiar, a fundamental psychological process that helps us to function. If the ridges on your cornflakes struck you in the same way as a sublime view of the Alps, then you might struggle to move on with your day. This does not mean the texture of your morning cereal is actually less awe-inspiring than the Matterhorn, however, as difficult as this may be to see in our everyday mode of consciousness.

When we see reality itself as being more fundamental than our ideas about it, we can feel it to have a freshness and aliveness that cannot be put into words, and we can be left with a sense that existence itself is deeply mysterious. When we drop our conceptual divisions, we can also be struck by the unity and interconnectedness of all that exists, kindling a sense of awe. Immanuel Kant coined the term *the sublime* for experiences of the vast power of nature that inspire an exhilarating wonder in us at being confronted by how small and powerless we are in comparison. If we consider this reverence for the vast power of nature in addition to the sense of fresh aliveness, mysteriousness, and awe at a reality that transcends our ability to fully encompass it with our minds, then we can see why people reach for terms like *the divine* to describe such experiences. After all, *divine* just means *relating to God*, and if we employ the naturalistic idea of God as the concept that points to that which transcends concepts, the nonconceptual totality of existence, then we can put talk of both God and the divine on a naturalistic footing.

This talk of the divine is useful because it is a common concept in the worldviews of many cultures, if not most. Two broad perspectives exist on the issue of where the divine is located in existence. Some hold the divine to be transcendent, existing apart from the world around us. Others believe it is imminent, being expressed in and through everything that exists. These claims are also tied up with the issue of where consciousness is felt to be located. For some, divinity is something that has its own independent, objective existence and is thus largely divorced from our experience of being in the world. For others, divinity is something to be experienced within ourselves through our own consciousness. In this way, divinity and the ability to experience are intimately tied together in many worldviews. Animistic cultures typically take the imminent view, with the divine permeating the natural world and being accessible to each of us, while contemporary monotheistic religions routinely position the divine as being separate from the earth in a transcendent heaven. The world that gave birth to modern science had drained the divine out of nature and locked it away in a place that only the priests supposedly had access to.

While animistic cultures still exist today, the initial transition away from animism in Eurasia appears to have begun with the development of agriculture around 10,000 BC. As humans settled in specific places and began cultivating crops, their material relationship with nature began to change. Humans developed greater control over their environment, and with it, they seemingly also developed a narrative of disconnection from and superiority to the rest of nature. In a hunter-gatherer society, one might respect the wild spirit of the untamed bull, whose sharp horns make it a worthy match for the hunter's spear, and may feel in one's bones the circularity of the mutual relationship between human and nonhuman: You eat the bull's flesh today knowing that one day you will return to the earth, becoming the grass that the wild cattle will in turn feed on. Regarding cattle as respected equals is more difficult once you have bound it in service to work the field. Here, the bull has gone from embodying the sublime power of wild nature and has been brought into the world of the familiar and mundane. It comes to be used as a tool to benefit the human, who is the dominant player in the relationship.

Along with this it makes sense that there would be a corresponding withdrawal of empathy for the subjective states of the animal. It would be too psychologically stressful to dominate a being that you felt to be your kin, day in, day out. The bull is now not only being treated as an object, related to in terms of its instrumental value for achieving the goals of the human, but it is also *seen* as an object. The human desire to control nature in order to insulate our species from the harms that can arise from the natural world leads to us turning subjects into objects in our own minds and simultaneously treating them as objects to benefit us. With the advent of agriculture and storable food sources, we see the ability to hoard wealth and power, leading to the domination and objectification of other humans, in addition to the nonhuman world. Along with the emergence of hierarchies of this kind, we see the emergence of divine rulers. Over much of the world, the special position of the ruler was justified in terms of the divine right of kings, who were given their blessing by a priestly class.

In some animistic cultures, shamans perform the priestly role of mediating with the divine. In an imminent conception of the divine, the shaman mediates between the powerful forces of the natural world and the human sphere. One might say that, by entering altered states of consciousness and transcending the rational mind, they bring their intuition to bear to see patterns that were not visible to the mind when in ordinary consciousness. Whereas the shaman mediates between an imminent divinity that is found in all of nature and is therefore in theory accessible to anybody, the hierarchy of power that a ruler sits atop requires a different narrative. Where there is a divine right of kings, there must also be a sense of the transcendence of divinity and a priestly class that functions as gatekeepers. In such cultures, God is no longer to be found in the body of the universe but is relegated to some ethereal heaven, set apart from the lowly world. If one wants divine wisdom of the forces that shape your fate, then you have to go through the priests and the ruler they serve to gain access to the divine.

We see the emergence of human exceptionalism and the divinity of rulers in ancient Mesopotamia, ancient Egypt, and the Mandate of Heaven in China. Interestingly, this does not seem to have been an immediate binary shift. Both transcendent and imminent views of the divine with their corresponding links to hierarchy and decentralization have existed alongside each other for a long time. In ancient Egypt, the gods continued to take the form of therianthropes of the kind depicted in paleolithic cave art, suggesting a connection to divinity in the natural world alongside the hierarchical access to the transcendent that was possessed by the pharaoh. In the Axial Age, we see Plato developing the superlative transcendent view of the world, in which what is most real is a transcendent realm of pure form. Aristotle, though, combined a sense of subjectivity pervading nature with human superiority. In his hierarchy of souls, Aristotle conceived of all living things, including plants, as possessing a vegetative soul that allows the organism to reproduce and grow, while nonhuman animals possess an additional sensitive soul that allows sensation and action, and humans also possess a rational soul that enables thought.

Whereas Eastern thought remained in touch with imminent spirituality, as expressed in such systems as Taoism, Buddhism, and Advaita Vedanta, in the West the transcendent vision of divinity would come to dominate. The Hebrew Bible, incorporated as the Old Testament by Christianity, clearly states that humans were made in the image of God to have dominion over the creatures of the earth: "And God said, 'Let us make man in our image, after our likeness: and let them have dominion over the fish of the sea, and over the fowl of the air, and over the cattle, and over all the earth, and over every creeping thing that creepeth upon the earth.'"[9] Christianity would come to employ Plato's philosophy in crafting its image of a transcendent heaven set apart from the world. This would culminate in the doctrine of Contemptus mundi, or "contempt of the world."

In reading the earliest texts that report the sayings of Jesus, such as those found in the Nag Hammadi library, we find what appears to be teachings about the imminence of the divine. We see this in the Gospel of Thomas:

> Jesus said, "If those who lead you say to you, 'See, the kingdom is in the sky,' then the birds of the sky will precede you. If they say to you, 'It is in the sea,' then the fish will precede you. Rather, the kingdom is inside of you, and it is outside of you. When you come to know yourselves, then you will become known, and you will realize that it is you who are the sons of the living father."[10]

The idea that the "kingdom of God is within you" even made it into the canonical Gospels that received the stamp of approval from the deeply hierarchical Roman Empire. It is perhaps unsurprising, however, that the overall message was transmuted into one of transcendence that justified the hierarchies of the church and state.

The mystical experience has the power to put us in touch with a vision of a decentralized, imminent spirituality that need not be mediated through the hierarchical structures of organized religion. As these insights are directly available in all times and places, we see them recurring again and again even within the Western trajectory toward transcendence and the gatekeeping of God by those in power. In ancient Greece and Rome, the Stoics, such as Seneca, held a pantheistic vision of the world, as did the philosopher Cicero. As late as the Renaissance, Italian philosopher Giovanni Pico della Mirandola wrote, "All this great body of the world is a soul, full of intellect and of God, who fills it within and without and vivifies the All."[11] We see it again with Spinoza's pantheism. It may be that this kind of worldview is not optional if we want to perceive our place in nature accurately.

ECOLOGY AND SPIRITUALITY

When we try to pick out anything by itself, we find it hitched to everything else in the universe. —JOHN MUIR[12]

Many of us go about our lives without paying much attention to the natural world. We might spend our days in cities with little nature, our thoughts preoccupied with the realm of human affairs. At other times, we may find ourselves out in the wild, connecting with the present moment. On such occasions, our minds can shift into a state of ecological consciousness. This is a way of seeing in which we apprehend beauty and value throughout the natural world, not just in ourselves. We may feel a sense of respect and awe at an insect or a flower. We may feel the distance between ourselves and the rest of the natural world falling away. There can be a sense of kinship with the plants and animals you observe, of recognizing yourself as part of nature. This sense of connection can grow to the point where you lose your sense of being something separate entirely, and you perceive yourself to be part of the vast ecological network that is the biosphere.

Understanding the place of the mind in nature is a tricky business, as our engagement with nature shapes our mind, and in turn our relationship with nature is shaped by our mental attitudes toward it. In the 1960s, interest emerged in this intersection between ecology and psychology, and by the 1990s a field of research had developed known as ecopsychology. The study of this intersection stretches back far beyond the origins of this particular discipline, however, and can be found far and wide, with Taoism and Zen Buddhism being particularly vivid examples. The clear seeing of the expansive circumstances that give rise to us through the process of nature is adjacent to mystical experiences of seeing one's perceived separation from the rest of the world as illusory. As a result, ecology and spirituality are intimately linked.

The man who coined the term *ecology* was a perfect exemplar of this connection. Ernst Haeckel was a contemporary of Darwin and an advocate for monism.[13] Haeckel used the term *monism* to point to the unitive, nondual, undivided nature of reality that is described in the world's religious traditions rather than simply the metaphysical position that only one kind of thing exists. In 1892 he published *Monism as Connecting Religion and Science*, in which he argued, "The monistic idea of God, which alone is compatible with our present knowledge of nature, recognizes the divine spirit in all things. It can never recognize in God a 'personal being,' or, in other words, an individual of limited extension in space, or even of human form. God is everywhere."[14] Haeckel was a pantheist in the mold of Spinoza. Lest one be inclined to celebrate Haeckel uncritically, however, it is important to note that he was also a eugenicist and proponent of the ideology of "scientific racism."

Ecological awareness increased dramatically in the 1960s, with landmark books in this area such as Rachel Carson's *Silent Spring* even being read by then US president John F. Kennedy.[15] In the 1970s, the mainstream narrative around our strained relationship with nature that had emerged over previous years was not radical enough for some. Rather than capturing the relevant truth that we are intimately part of nature, much of the environmental movement appeared to be operating in a worldview in which humans were still seen as separate from the natural world, which functions as a container for us, a stage on which the superior humans conduct their business. Even the term *environment* points to this worldview of nature as a context for us rather than something that we participate in as part of it.

A Norwegian ecologist by the name of Arne Næss sought to differentiate the more radical view of our place in nature from this shallow vision
of ecology. In the 1970s, he coined the term *deep ecology* for the perspective
in which we are understood to be full participants in the natural world.[16] As
with the thought of Haeckel, deep ecology embraces the expansive states of
consciousness in which we see ourselves as not truly separate from the rest
of reality. In the case of Næss, he explicitly positioned himself in the pantheistic tradition of Spinoza, writing a paper titled "Spinoza and Ecology"
in the 1970s that connected the work of this philosopher with thought in
the emerging ecological movement.[17]

Multiple thinkers over the distance of centuries have seen a connection
between expansive, mystical states of consciousness and ecological awareness. In recent years, we have begun seeing hard evidence of the psychological reality of this connection. Researchers at Imperial College London's
Center for Psychedelic Science conducted a study in which they sought
to measure a psychological trait known as "nature relatedness" before and
after participants received a high dose of psilocybin.[18] Nature relatedness
captures the experience one has of feeling connected to the rest of the natural world. They found that nature relatedness did indeed increase after the
psychedelic experience. A participant in another psilocybin study reported,
"Before I enjoyed nature, now I feel part of it. Before I was looking at it as
a thing, like TV or a painting. [But] you're part of it, there's no separation
or distinction, you *are* it."[19] The capacity for empathy alone does not tell us
anything about whether that empathy is valid; there is always the risk of
anthropomorphization, of projecting traits that we possess onto things that
do not possess them in reality. What this does show, however, is that there is
a spectrum from feeling separate from nature to feeling fully part of nature.
What we can take from this is that the widespread feeling of being unlike
the rest of the living world cannot be taken as simply self-evident; the notion that there is a commonsense consensus on this issue is an illusion.

The biophilia that can seemingly be produced by the psychedelic experience is no transient effect. The increased ecological consciousness that
is seen after such experiences has been found to persist and even to be
increased two years after the experience.[20] It is also associated with tangible pro-environmental behaviors and a shift from seeing the natural
world around us as something to be owned and exploited to something

that should be respectfully stewarded. The connection between psychedelic experiences and nature relatedness is so strong that Richard Doyle, a theorist in this area, has proposed that psychedelics instead be called ecodelics.[21]

In the study described earlier, the stronger the experience of ego dissolution a participant underwent, the stronger the connection to nature that was felt after the experience. Our ego, or our sense of self, mediates our sense of feeling separate from the rest of nature. When its boundaries are dissolved, we're left with the feeling that, beyond our imaginary borders, there is only the vast web of nature, a fabric of connections rather than separate individuals. When the illusory division in consciousness between subject and object is seen through, the mind's true undivided character is revealed; the contents of consciousness are intrinsically illuminated, without the need for a separate subject to peer at and inspect them from a distance. The mind in this state is seen to be incredibly simple, consisting only of appearances arising in awareness. We could also say arising *as* awareness, as there is no true division between consciousness and its contents, in the same way that Monet's *Water Lilies* really are just paint on a canvas—they are not separate from the medium in which they appear. When consciousness is seen to be not something complex but instead to be simple feeling, the possibility that consciousness may exist in systems that do not possess big, impressive brains becomes deeply intuitive. This opens up the consideration of perspectives beyond the nervous-system-centered approach for consideration when it comes to understanding where consciousness emerged.

SETTING THE STAGE FOR EXPLAINING CONSCIOUSNESS

The most beautiful thing we can experience is the mysterious. It is the source of all art and science. He to whom this emotion is a stranger, who can no longer pause to wonder and stand rapt in awe, is as good as dead; his eyes are closed.—ALBERT EINSTEIN[22]

When we consider consciousness within Descartes's worldview, in which nature consists of unfeeling mechanisms, we are unable to make sense of experience, unable to understand how minds could possibly have emerged from unconscious matter. The first part of this book critically examines where this worldview goes wrong and points to another way of seeing the world, a vision of reality that is usually associated with spirituality and certain religious traditions yet is compatible with science. Our commonsense impression of consciousness is that it consists of being a thinking subject, of being the "me" that has thoughts and experiences, a me that is separate from the rest of the world. In "spiritual" modes of consciousness, such as the mystical state, this separation breaks down. There can be the recognition that the fundamental nature of consciousness is like a bare canvas in which all experience arises, both the experiences that we usually think of as arising from inside and constituting ourselves and those that usually seem to arise from outside us. When this duality breaks down, there can be a sense that reality is fundamentally undivided, that our sense of being a separate self is an illusion. If we take this insight seriously, then we are on good footing to understand the place of consciousness in the natural world.

The mechanistic story of nature disenchants the world around us, giving us the impression that it is all understood and under our control. When we experience consciousness beyond conceptual divisions, we can be struck instead by an irreducible sense of mystery and awe at existence. Without feeling comfortable with the sense of mystery that arises from seeing the nonconceptual nature of existence, we are liable to mistake the source of our awe. Today, many scientists feel that, consciousness aside, the natural world is thoroughly understood and devoid of mystery. This is a case of mistaking the map for the territory, of believing the actual world around us to be the same thing as our abstract models of it. The one phenomenon that doesn't fit into this way of thinking, however, is consciousness. The fact that you really do feel pain when you are injured or feel love for someone you care for cannot be abstracted away. Experience feels almost like a magic trick from this perspective, an unfathomable mystery that confronts us in every moment. Here, what is perceived as the mystery of consciousness is actually the mystery of being in disguise, the sublime fact that existence cannot be contained within our ideas of it. This imminent spiritual sensibility, this awe at existence, is a necessary part of intuiting the place of consciousness in the natural world. The solution is to lean into the mystery, to permit the contemplative's appreciation for the nonconceptual alongside the conceptual play of science. This all may seem quite distant and hard to understand, but it is in reality closer than close. All it takes to get familiar with it is to begin observing one's experience, as in any form of meditation.

We have come a long way from the typical scientific worldview, but I believe this updated version is both naturalistic and necessary if we are to understand consciousness. I have questioned the reality of self, separation, and substance. I have considered a process-relational vision of the world in which matter and mind are two modes of operation of reality, à la Spinoza's neutral monism. I've shown that, while idealism and panpsychism may capture the groundedness of the formless awareness at the core of experience, as well as the continuity of mind with matter, the physicalist understanding of the dependence of experience on physical structure is also necessary in order to explain consciousness. I have also shown, however, that the assumption of human exceptionalism can be understood as a biased belief rather than a fact. How are we to think about consciousness and its relationship to the brain, the wider nervous system, and the bodies of living things? This is the relationship that must be explained in order to truly close the gap between matter and mind, and it is what part II examines, now that the philosophical foundations for such an understanding have been laid.

OUTSIDE IN

The Science of Consciousness

EIGHT

CONSCIOUSNESS AND THE BRAIN

The Brain—is wider than the Sky—
For—put them side by side—
The one the other will contain
With ease—and you—beside
—EMILY DICKINSON[1]

Somehow, we feel, the water of the physical brain is turned into the wine
of consciousness.—COLIN McGINN[2]

EXPLAINING THE EMERGENCE OF CONSCIOUSNESS

As a teenager, I began having strange tingling sensations on the back of my head. I had had an injury near my eye shortly before these symptoms emerged, and my doctor sent me for an MRI to check whether any damage may have arisen as a result. My doctor's concern arose from the fact that the retina of the eye is part of the central nervous system, the rest of which is hidden away within the skull and spine. When an ophthalmologist looks into your eyes with their ophthalmoscope, they are seeing a part of your brain. I remember nervously receiving the results of the scan and wondering if it might reveal a blank cavity where a brain should be. Fortunately, everything was fine, and several years later I was very happy to have in my possession images of my own brain while I was studying this very organ at university. Looking at these images now, all that is contained in my skull appears to be a lumpy, folded mass of ordinary-looking tissue, with ghoulish Rorschach-ink-blot faces emerging as the cross sections march forward through my head. Supposedly this wrinkled organ had been responsible for all the feelings I had ever had, every taste, every smell, every thought and desire, my dreams and imaginings, and my hopes for the future.

We are confidently told of this connection by many scientists, who claim that this organ alone holds the secrets to our mental lives. Such is the certainty of these voices that I never once doubted that the brain was the place to look if I was interested in understanding the origins of consciousness. Francis Crick wrote in his book *The Astonishing Hypothesis*, "A person's mental activities are entirely due to the behavior of nerve cells, glial cells, and the atoms, ions, and molecules that make them up and influence them."[3] "Entirely due" to the cells of the brain alone is a very strong statement that suggests that nothing else whatsoever contributes to our mental lives, and this statement was made by the man who arguably founded the field of consciousness science. I believe it is this assumption that has prevented us from seeing the true origins of consciousness, which lie outside the brain. I do not say this lightly. As a neuroscientist I have spent my career believing that the brain held the answers to the question of how consciousness came into existence. This does not mean that the brain is not involved in conscious experience in our species, however. Before we move beyond the brain-centric paradigm, let us explore the fascinating work that has been conducted over the last several decades by pioneering scientists who have explored the connection between consciousness and the human brain. The findings of this research will serve us well in understanding what physical processes the evidence suggests are associated with consciousness.

THE HIDDEN BRAIN

As an undergraduate I was once told a story by my tutor, a neuropsychological researcher, about a patient she had worked with. This patient had received damage to an area at the back of the brain on the right hemisphere called the parietal lobe, producing a condition called hemispatial neglect. In this condition, the left side of the world is either ignored or not experienced at all. Patients will eat the food from only the right side of their plate, for example, and then if the plate is rotated, it appears to them as if it has miraculously refilled. Ask them to draw a clock face, and they can draw a circle, but all the numbers will be crammed into the right-hand side. One day she was wheeling this patient in a wheelchair, and she turned to take them into a room. As she began to turn the wheelchair, the patient stuck their leg in the way to stop their forward progress. This kept happening until she realized she was turning left through the doorway. From the patient's perspective, they may have felt they were being turned directly into an abyss of nonexperience.

What could it be about this lump of meat in our skulls that could allow it to experience things as ghostly and immaterial as ideas or hopes? The human brain is astoundingly complex, and many look to this complexity for an answer to this question. From the outside it looks like a lumpy, gelatinous mass, a ball of approximately six inches across that weighs only about three pounds. Its fine structure, however, consists of approximately 100 billion neurons, roughly the number of stars in our galaxy. The quantity of connections between these brain cells is even more staggering, numbering in the trillions. This incredible organ manages to achieve things that even our most powerful technology cannot, and it does so with a power consumption comparable to that of a light bulb.

The technological analogy here is apt, as for the last half-century, the computer analogy of the brain has dominated the field. In my first term as an undergraduate at Oxford studying the brain, I was told in a lecture that the brain-computer analogy is only the most recent in a series of flawed comparisons between the brain and whatever is the most exciting technology of the day. Surely this statement couldn't be correct, I thought. It seemed obviously wrong to me to think of the brain as anything like a steam engine, as some scientists had evidently suggested in the past. But the idea that it might be a general problem solver like a computer seemed eminently reasonable. Our ancestors may not have had the vision to see beyond their flawed analogies, but surely we could. As I show later, it turns out we are not so different from our predecessors.

CONSCIOUSNESS AND THE BRAIN

If I asked you why you have a brain, what answer would you give? You might say, "To think," or if you were familiar with the brain-computer analogy, then you might say, "To process information." Yet even in this framework, this is not an end in itself. The ultimate goal of the brain's activity is to allow you to move effectively in the world. The brain, after all, was sculpted by evolution to allow you to survive, to find delicious berries, and to avoid fearsome predators. The nervous system seamlessly connects sensation to action, continually orchestrating our interactions with the world around us in every moment. This process begins with our sensory receptors, the retina of the eye, the cochlea of the ear, the skin, and the olfactory receptors of the nose, among others. These receptors convert stimuli, from the touch of a gentle breeze on the skin to the photons of a sunset, into electrical signals that the brain can interpret. These signals then embark on a journey through a labyrinth of neural pathways.

As the signals navigate this neural network, they pass through a central processing hub of the brain called the thalamus, a structure located in the very middle of the brain, and ascend toward the cerebral cortex (see figure 8.1). The cortex is a sheet of neural tissue about the size of a tea towel, although rather than appearing smooth and flat, it is scrunched and folded to fit inside the skull—it is the outer part of the brain, which most people picture when they think of a brain. This structure contains neural circuits that are capable of performing a wide range of functions, from vision and hearing to decision making and movement. Sensory information in the form of electrical impulses is initially processed in regions of the cortex that lie at the back of the head. This activity then radiates forward, with increasingly sophisticated functions being implemented in the neural circuits it courses through on its way to the front of the brain.

In the frontal areas behind the forehead, this information is used to guide behavior, sending signals down the spinal cord to control the muscles of the body, from the graceful dance of a pianist's fingers to the explosive power of an athlete on the field. Muscles contract, limbs move, and the body responds with precision to the brain's commands, translating sensation into tangible action. In every moment of the day, the brain keeps this loop from sensation to action running, where sensory information flows inward, perceptions take shape, cognition guides decisions, and actions ripple outward, allowing you to have an impact on the world. This is only one way of thinking about how the brain functions, however, and I show this loop from sensation to action from another perspective in chapter 10. For now, what is important is that the brain manages to connect sensation to action so that input from our senses can guide our behavior.

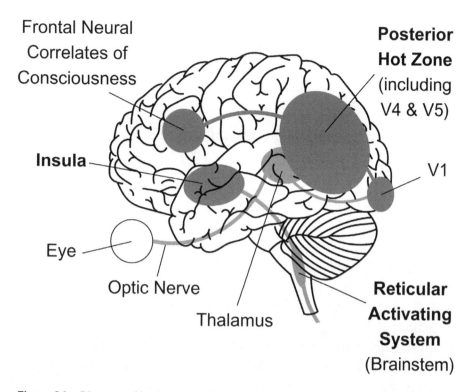

Figure 8.1. Diagram of brain areas relevant to consciousness. Darker circles and ovals correspond to areas of the cortex, the outermost structure of the brain (V1, posterior hot zone, frontal neural correlates of consciousness, insula), while lighter ones correspond to subcortical structures that are located deeper within the brain (the thalamus, the reticular activating system [RAS] of the brainstem). The insula is positioned in a unique location among these areas, as it is part of the outermost part of the brain in anatomical terms, the cortex, but is hidden from our direct view underneath another lobe of the cortex. In bold are the core brain areas implicated in producing conscious content, the posterior hot zone for vision and both the reticular activating system and insula for bodily feeling. Gray lines show major connection pathways between brain areas. *This image has been designed using assets from Freepik.com, https://www.freepik.com/free-psd/brain-outline-illustration_65105163.htm.*

Where in this process does consciousness come in? If we look at the individual neurons that make up the brain, we see nothing apart from physical processes taking place—we see no special properties that should make them capable of generating something seemingly nonphysical like experience. If we zoom out from the microscopic cellular level and instead look at brain areas that are on the order of inches in size, areas that consist of networks of millions of neurons, then we can see which brain regions appear to be connected to consciousness. Perhaps something special occurs at the level of neural networks of this kind that could account for the seeming miracle of experience. I consider this possibility in the next chapter. For now, let's explore which brain areas have been implicated in human consciousness.

VISUAL CONSCIOUSNESS

Let us take the case of vision and follow the sensory signals that contribute to experience along their journey through the brain. The process of vision begins with the lens of the eye focusing an image on the retina. The optic nerve then carries these signals from the retina to the thalamus and then to the cortex. The first cortical area that the visual signals arrive at, located at the very back of the head, is called V1, short for the sensible name visual area 1.

Following the Russo-Japanese War of 1904–1905, Japanese soldiers came back with damage to this area of the brain and a corresponding loss of vision, despite their eyes being undamaged, that gave the first clue as to its function.[4] Bullets at the time did not explode on impact as they typically do today, leading to many passing relatively cleanly through the heads of soldiers, remarkably sometimes allowing them to survive. Typically, these gunshot survivors did not lose vision in their entire visual field but only in circumscribed regions. This allowed researchers to map the relationship between the brain areas damaged and the regions of visual experience that were lost as a result. What this revealed is that the visual space in front of us is reflected in the organization of this brain structure.

The spatial arrangement that exists on your retina is preserved as signals course through the optic nerve and thalamus to arrive at this brain region, creating a map of visual space. This map is reversed with respect to the space in front of your eyes, however, so that the left hemisphere of the brain carries signals that originate from the right portion of the visual field, while the right hemisphere carries signals from the left visual field. Each visual brain area comes as a pair, with the same region existing in each hemisphere of the brain and processing the signals from the opposite side of visual space.

George Riddoch, a neurologist, noticed a strange aspect of this form of blindness.[5] Some of his patients knew when an object moved within the region of the visual field in which they were blind, even though they had no conscious experience of the object's properties, such as its shape or color. Many years later, in the 1970s, the neuroscientist Larry Weiskrantz would observe a similar phenomenon that he termed *blindsight*.[6] Weiskrantz would present a visual stimulus in the blind region of his subject's visual field and would ask questions about it, such as its location or appearance. With no visual experience, the subject would feel as if they were guessing when they offered a response. Despite their lack of awareness of what was in front of them, Weiskrantz found that his subjects with blindsight could successfully answer these questions at a level above chance. Similar successful performance was found for visually guided behaviors, such as posting a letter through a slit of different orientations, while no visual awareness was present. This suggested that, in some situations, conscious experience was not necessary for behavior, and if the physical basis of consciousness were to be found anywhere, then it might be found in the cortex.

Reptiles do not have a cerebral cortex like ours. They instead process visual information in a structure called the optic tectum, a brain area that allows for the visual guidance of movement. Mammals like us possess a corresponding structure, called the superior colliculus. The superior colliculus receives input from the retina and processes visual signals in parallel with the thalamus and cortex. It is thought that this evolutionarily older pathway might be responsible for mediating the behaviors that can be performed by patients with blindsight, seemingly without conscious experience. From this, it seems that the superior colliculus is capable of guiding behavior in an unconscious way, while visual processing in the cortex appears to be associated with consciousness, given that visual experience is lost with damage to these visual cortical areas. This is one reason that many researchers believe that there is something about the cortex, something absent from structures like the superior colliculus, that is responsible for the generation of consciousness.

In the decades that followed, the precise role of V1 in processing visual signals began to be elucidated. This area identifies the basic components of the image, mainly the oriented lines that make up the edges of objects.[7] Consider how it is that you can identify a circle with two dots and a line inside as a face. This is possible because the brain picks out the transition point from the object to the background, which we call the edge, and emphasizes such transition points in its processing. The result is that a 2D circle approximates the way the brain represents a 3D spherical object. This is an elegant way to send signals, as signaling the visual information at each location would lead to a high level of redundancy.

Imagine the image of a white piece of paper on a black surface. One way to transmit this information would be to signal every black location and ignore every white location, effectively using a binary black-versus-white sensor to detect what is present at every single part of the image. Given that there are so many black locations next to each other, however, there is no need to signal the same information again and again. Rather than employing this laborious strategy, it would be more effective to assume no change until a change occurs. In this mode, we can envisage change detectors being used to signal the information rather than simple black-and-white sensors. This way you only have to signal the beginning of the black region once and the transition to the white region once, saving energy, which is crucial in evolutionary terms. The brain is the most energetically expensive organ in the body, consuming about 20 percent of the body's energy despite making up approximately 2 percent of its weight, so energy conservation is particularly important for this organ.

Let us take a moment to look at the mechanism that is thought to generate this visual code in the brain. A surprisingly simple and elegant circuit is capable of functioning as a change detector over visual space. If one neuron responds positively to either black or white in one region and another responds negatively to the same stimulus in the surrounding region, then the signal will balance and cancel when there is no change in that part of visual space. When there is an edge, however, a transition from black to white, then there will typically be an imbalance, leading to a net positive or negative signal.

We can see from this that straightforward mechanisms implemented by relatively simple neural circuits are capable of performing sophisticated visual processing, seemingly leaving no need for consciousness. This is the core issue at play when looking for the origin of consciousness in the brain. The brain appears to function perfectly well with physical nonconscious mechanisms, so what could ephemeral experiences contribute to the functionality of this organ? If an evolutionarily old structure like the superior colliculus can seemingly function perfectly well without consciousness and animals that don't possess a cortex like ours, such as reptiles, can perform visual processing with such brain structures, then what does experience add? Why do our brains not simply process visual information with a highly developed version of blindsight in which no consciousness is present? Why does this processing go along with the visual experience of sight?

The information that is signaled in V1 consists largely of edges, as discussed earlier. Not all this information appears to make it through into consciousness, however. If one is exposed to a set of tilted lines for long enough, then the neurons that signal their orientation will adapt, leading to an aftereffect where a new stimulus is experienced as tilting in the opposite direction. It is also possible to show tilted lines that are so fine that they blend into a homogenous surface in consciousness. Despite not perceiving separate lines of a particular orientation, the viewer will still experience the orientation-dependent aftereffect.[8] This suggests that the orientation information is present in V1, something that must be the case if it is to produce the aftereffect, but that these signals do not consistently contribute to conscious experience, as the lines and their orientation are not consciously perceived. This further suggests that V1 is not the area of the brain where information is made conscious. We must therefore look further along in the visual pathway if we are to find a specific brain mechanism that could generate experience.

After V1, the visual information is sent to multiple other visual areas in the cortex. These have names like V4 and V5, but processing does not occur sequentially through each of these areas; instead they all process the information in parallel. V1 lies at the very back of the head, and these higher-visual areas lie slightly in front of V1. V4 is involved in color processing, while a nearby area of the brain, known as the fusiform face area, is involved in recognizing faces. V5, meanwhile, is involved in processing visual motion. Damage to each of these areas results in a corresponding lack of conscious experience in the domain that they appear to be involved in.

When damage occurs to V4 as a result of a tumor, stroke, or physical trauma, for example, patients are typically left with a condition called cerebral achromatopsia.[9] Individuals with this condition lack perception of color in their visual experience and instead experience the world as monochrome. This provides convincing evidence for the role of V4 in the conscious experience of color. Further evidence comes from the fact that, if V4 is only damaged on one side of the brain, then only the opposite visual field loses its color. Damage to V5, the motion-processing area, produces a corresponding loss in the conscious perception of motion, known as akinetopsia.[10] Patients with this form of motion-blindness report seeing the world as a series of images with no intervening motion. As they cross the road, they may see a car at one distance, then closer, then closer still, and must infer that it is moving toward them rather than consciously perceiving it directly.

We can see from the effects of these different patterns of brain damage that different parts of the cortex contribute to different aspects of visual perception. This seems to dispel hopes of finding a single "consciousness center" in the brain that is responsible for generating all experience. These areas are broadly clustered in an area just above and behind your ears on both sides, toward the back of the brain. As a result of this anatomical location, this broad sweep of brain areas, including V4 and V5, among others, has been termed the *posterior hot zone* for conscious experience. Further evidence for the role of the posterior hot zone in consciousness comes from brain-imaging studies, which show the activity of these areas to correlate with changes in perception.[11]

In the first chapter, I explained that one way to map to neural correlates of visual consciousness is through the binocular rivalry paradigm, where a researcher presents different images to each eye of an individual. Rather than perceiving a combination of the two images, only one is perceived at a time, but which one is perceived switches back and forth over time. In this situation nothing changes in the physical setup of the images, nothing changes in the input being received by the brain, yet conscious experience does change. This creates an ideal opportunity to track what activity in the brain might be associated with conscious experience itself. Studies of these kind confirm the presence of activity in the posterior hot zone during conscious experience but also reveal activity in areas in the front of the brain. Might these areas also be involved in generating consciousness?

THE VIEW FROM THE TOP

The perspective we've explored so far looks at how consciousness might naturally bubble up as a result of sensory processing. Not all the sensory information that is processed by the brain appears to enter consciousness, however. Perhaps there's another way of thinking about experience that might make sense of this. Consider your basic intuitions about what happens when you become conscious of something. Many people feel that this process involves them mentally inspecting the thing that they are conscious of in an active way rather than sensory processing just passively volunteering things for them to be conscious of. The word *subjectivity* implies something of this kind, a subject that claims ownership over experience and can choose what to become aware of.

I explore in earlier chapters how there isn't a single indivisible "thing" in the brain that could be called a self or a subject, but we can imagine a brain network that might inspect other brain processes on behalf of the organism as a whole. One class of theories, known as higher-order theories, holds that the contents of a brain process become conscious when they are represented by higher-order networks in the brain that are located further up the hierarchy of processing. Another influential theory, called global workspace theory (GWT), proposes a slightly different mechanism, one of broadcasting the information that is to be made conscious throughout the brain so that many different systems have access to it.[12] What I describe here are mechanisms that underpin cognitive processes. Such "executive functions" are performed by regions in the front half of the brain, such as the prefrontal cortex. If consciousness involves such mechanisms accessing the contents of other brain processes, thereby allowing "us," the subject, to become conscious of their contents, then we would expect the neural correlates of consciousness to be found in the front of the brain. We can call such theories subjective access theories.

As I showed earlier, experiments using the binocular rivalry paradigm have found that there are correlates of consciousness in the front of the brain, in addition to areas in the posterior hot zone. One issue with these findings, however, is that, for scientists to collect data about a subject's conscious state, the subject must convey to the experimenter what they are experiencing. This could be done verbally or by pressing a button when the image switches in their perception, for example, but either way they would have to use certain brain networks to communicate their experience. The networks that are involved in such communication happen to be the frontal networks that had previously been considered a neural correlate of consciousness, in support of the subjective access theories. This makes it difficult, however, to determine whether these areas really do contribute to consciousness or whether they just report on what is consciously experienced. The issue is whether the ability to introspect about our conscious experience plays a role in generating the experience itself or if it merely reflects the act of reporting on experiences that are already conscious.

One ingenious study in 2014 succeeded in teasing apart these issues. In this study, researchers sidestepped the problem of subjects intentionally communicating their experience by showing that reflexive eye movements, such as changes in pupil size, mirror the changes in conscious perception.[13] By monitoring these eye movements, the researchers could determine what the subject was experiencing without them having to report anything. The experimenters found that, when the experiment was set up in this way, the activity of frontal areas diminished, indicating that their role is primarily in the reporting of conscious contents rather than the generation of these contents. What remained, however, was activation at the back of the brain in the posterior hot zone.

FROM COGNITION TO FEELING

Given the convincing evidence from multiple sources for the relevance of the posterior hot zone to consciousness, can we now call off the search for the brain basis of consciousness? Not quite. These findings are typically confined to visual experiences, and while we are a very visual species, we do have conscious experiences of other kinds. As these correlates were found largely in visual parts of the brain, it appears unlikely that they mediate experiences in all modalities, and the research bears this out.

What is the simplest and most primordial kind of experience we can have? Before we developed sophisticated eyes that could focus an image on a retina or ears that could pick out fine details in the frequency content of a sound, our ancestors had to feel the internal state of their own bodies to stay alive. In order to survive, we must keep our body temperature in a very narrow range, and the same is true for other physiological processes. Without the ability to sense the state of one's own body, there is no way to regulate it, which would lead very quickly to death. Perhaps it was this evolutionary imperative that led to conscious experience evolving, first in the form of bodily feeling.

Signals arising from inside the body are known as interoceptive signals, and they arrive to the brain through the brainstem, located at the top of the spine, before passing to the thalamus and an area called the posterior insular cortex. This is the back portion of a piece of neural tissue that is tucked away on the inside of the brain. If you look at the cortex from the side, it looks something like a boxing glove, with its two protrusions where the thumb and fingers go. The insula is like the bit of the glove located in the crook of the thumb, hidden away when the hand is formed into a fist. Nestled in this secretive position, the posterior insula processes these bodily signals and passes them on to the front half of the same brain region, the anterior insula. Neuroscientist Bud Craig has suggested that, in the anterior insula, these signals are re-represented, making them conscious.[14]

This story ties in nicely with a way of thinking about consciousness proposed by the influential Portuguese neuroscientist Antonio Damasio.[15] His perspective focuses on homeostasis, the ability of the body to regulate crucial parameters, such as body temperature, oxygen levels, salt concentrations, and so on, so that they all stay within their appropriate ranges for the body to function successfully. Damasio argues that consciousness arises when the brain makes an "image" of this regulatory process occurring in the body. He sees consciousness as originating with feelings, experiences of the state of the body that signal how well the process of homeostasis is going.

Brain scans have revealed a wide range of subjective experiences that are associated with activation of the anterior insula, providing evidence for the involvement of this brain area in consciousness.[16] These include feelings that arise from bodily sensations, such as temperature, thirst, breathing, sexual arousal, wine tasting, and having an overly full stomach or bladder. Other experiences connected to the body, such as the sense of body ownership, feeling bodily pain or empathic pain when witnessing a loved one in pain, and hearing music that one just learned to play, also activate this brain region. Enjoyment of music, too, activates this area. It is activated by emotions such as romantic love, maternal love, fear, sadness, anger, happiness, disgust, and uncertainty. Finally, it is also associated with elevated experiences of beauty, "union with God," and the subjective effects of the psychedelic brew ayahuasca. The feeling that we know something before we have recalled it, as when something is on the tip of one's tongue, also activates the anterior insula. This is only a partial list of the many subjective experiences connected with this brain region. By imagining what your inner world would be like if your insula were to become damaged, potentially impairing all these aspects of experience, you can get a feel for its significance in human consciousness. Given that the insula senses the internal state of the body, perhaps we should be looking deeper into the organism than the cortex for the origins of consciousness, to processes more fundamental to the operation of the living body.

THE BRAINSTEM AND LEVELS OF CONSCIOUSNESS

When considering which brain areas might produce conscious bodily feeling, I jumped over the brainstem, the first place in the brain that these signals are processed, in favor of focusing on the insula cortex. Might these evolutionarily more ancient brain structures play a role in consciousness? While the cortex appears to contribute to the content of visual experience, researchers also think about consciousness in terms of its level. Here some confusion can arise between two different ways in which we think about being conscious. The first is in terms of having experiences, as it has been used throughout this book, but we also talk about being conscious when we are referring to being awake. These two uses of the term *conscious* are by no means the same, as you can have experiences while asleep and dreaming. In such a case, you are unconscious in terms of wakefulness but conscious in terms of experience. Despite these complications, many people find it intuitive to speak of levels of conscious experience that correlate with levels of wakefulness, with more rich and vivid conscious experiences being had while wide awake and a lower level of consciousness being associated with drowsiness and sleep.

The brainstem contains a neural structure called the reticular activating system (RAS), which is responsible for our overall level of physiological arousal. The neurons in the reticular formation of the brainstem use a variety of substances, including well-known neurochemicals such as dopamine and serotonin, to modulate the activity of other areas of the nervous system. According to those who believe that the cortex is responsible for consciousness, the activity of the brainstem is a prerequisite for consciousness, but the cortex is necessary to bring consciousness into existence. According to the cortex-centered view, if we imagine consciousness as a TV show playing on a television, the brainstem is like the television's power supply, while the cortex is like the signal that produces the moving images—you need the power supply to be working in order to watch the TV show, but the power supply isn't responsible for generating the TV show. Others believe this picture does a disservice to the brainstem. They argue that this brain structure is far more central to the phenomenon of consciousness rather than being a mere power supply. Given that this structure signals important physiological information that forms the basis of bodily feeling and perhaps ultimately emotion, it has been argued by neuroscientists, such as Damasio and Mark Solms, that the brainstem contributes not only to conscious level but also to conscious content. They have suggested that this structure is the true locus of consciousness in the brain.[17]

The subjective experiences of embodied feeling that appear to be mediated by the brainstem and insula would not be possible without the living body that gives rise to interoceptive signals. Given the evidence for the involvement of these brain areas in consciousness, we can see a picture emerging in which the feeling body plays a crucial role in consciousness. In keeping with this, the origin of felt experiential states appears not to be in the most evolutionarily recent parts of the brain but in the most ancient, those involved in the regulation of the living body. Perhaps we should look beyond the brain if we want to understand the true origins of consciousness.

NINE

IS CONSCIOUSNESS A COMPUTATION?

A computational device is incapable of developing a mind. We got consciousness not just by being clever.—ROGER PENROSE[1]

SPECIAL STUFF

I have shown that science has done much to map the physical structures and processes in the brain that correlate with consciousness. What is it about such brain areas that seemingly give them the ability to conjure something as ephemeral as the experience of seeing a rainbow? Imagine making a machine that performs exactly the same functions as these brain areas: It takes in visual input from eye-like cameras and encodes the signals into representations that are then processed by the machine. Would this machine have a conscious experience? If so, then why? How do we get from this objective description of how the system operates to subjective feeling? How are we to make the leap from physical descriptions of a machine, or of the brain for that matter, to the conscious experience itself? What story could I tell you about electrical signals in nervous tissue that would give you a satisfying explanation of the feeling of being in love or the smell of freshly baked bread? At this point, we must turn back from the science to philosophy.

If we zoom in on such a brain area, we see a collection of biological cells. It is not only the electrochemically active neurons that are present but also other brain cells, ones that perform structural and immunological functions and affect the activity of neurons in multiple ways. If we focus on the neurons, we see that they constantly shuttle electrically charged atoms called ions across their membranes, depending on which molecules are present. This results in fluctuations of electrical activity in the neurons. Could there be something about the physical properties of brain cells or perhaps a physical property associated with them, such as electricity, that is responsible for consciousness? The other option is that consciousness has nothing to do with the material properties of the brain and its neurons but is instead a function that it performs. Let's consider the physical structure of the brain first.

IS CONSCIOUSNESS PHYSICAL?

Perhaps the atoms that make up the brain are responsible for its being conscious. This is the stance of panpsychism, but it doesn't explain why a brain area like V4 would be associated with color experience while other areas are not. If all atoms are conscious, then we might expect all brain areas, and all parts of the body for that matter, to be related to the contents of consciousness, but this does not appear to be the case. Panpsychism does nothing to solve the combination problem of how the experiences of atoms could aggregate to generate the experience associated with the organism as a whole.

What about electromagnetism, the physical force that underpins the electrical charge used by neurons to send signals? Considering the electromagnetic field as the basis of consciousness takes us away from particles and into the fields that govern their interactions. There have been multiple attempts to describe consciousness as being equivalent to electromagnetic fields, but this idea does not seem to account for how consciousness relates to the brain.[2] For example, if the brain is divided in two during surgery, then consciousness, too, appears to be divided into two islands of experience.[3]

In one study with patients who had undergone this procedure, a different picture was shown to each hemisphere of the brain. This is possible because, as we have seen, all input from the right side of the visual field is routed to the left hemisphere and vice versa for the other side. The right hemisphere was shown a snow scene, while the left hemisphere was shown a picture of a chicken's claw. The patients then were shown multiple other pictures and were asked to point to the ones that were associated with what they had just seen. The right hemisphere, which controls the left hand, correctly pointed to a snow shovel, while the left hemisphere, which controls the right hand, pointed to a picture of a chicken. The left hemisphere is the side that possesses the necessary brain structures for language, and so, when asked why they had pointed to a snow shovel in addition to a chicken after having only seen a chicken's claw, it invented a reason due to having no access to the conscious experience of the right hemisphere that had seen the snowy scene. "Oh, that's simple," the participant said. "The chicken claw goes with the chicken, and you need a shovel to clean out the chicken shed."[4]

In such cases, experience does indeed appear to be divided in two, but the overall electromagnetic field associated with the brain cannot be divided by a scalpel in this way. If consciousness were equivalent to this field, then it should remain unaffected by the surgery rather than being split in two. Instead of being synonymous with the electromagnetic field, the structure of consciousness appears to depend on the physical structure of the brain.

Perhaps it is the electrical characteristics of neurons in particular that underpin experience. Neurons are often presented as a unique type of cell in terms of their ability to carry an electrical charge, but this is not the case. Neurons do have very characteristic electrical dynamics, typically producing short pulses known as action potentials or spikes, but they are not unique in their capacity to carry electrical charge. In fact, all cells in the body are associated with a charge of this kind that is used to perform a variety of functions. The electrical properties of neurons alone cannot explain why they are associated with consciousness and other cells are not.

A related idea comes from anesthesiologist Stuart Hameroff and famed physicist Roger Penrose, who won the Nobel Prize for their work in physics. Their orchestrated objective reduction (Orch OR) theory holds that consciousness is produced by quantum events in structures called microtubules that can be found inside neurons.[5] The same microtubules are found in cells throughout the body, however, so their presence in the brain cannot be used to explain why the brain in particular is associated with consciousness in humans.

If consciousness is not due to a special physical property of individual neurons, then might it be due to the networked structure of the brain as a whole? Integrated information theory (IIT) holds that consciousness is synonymous with how integrated any system is.[6] According to IIT, consciousness is the way in which a system exists for itself. Something can only really be said to exist if it exerts some causal influence on something else in existence, and integrated information tries to capture the extent to which a system causally interacts with itself. In this framework, the rich interconnectivity of the brain would lead to it being highly conscious, but even a particle or a thermostat would be associated with some nonzero level of conscious experience. I come back to IIT later.

Another way of thinking about the relationship between the physical stuff of the brain and subjective conscious experience is to say that they simply are the same thing. I show later that thinking in terms of such "identity relationships" is crucial, but in this context, this claim is patently nonsensical. Imagine you and a friend are observing someone knitting a scarf. Your friend can't understand how it is that the scarf gets made through this process, but they can see that there is a connection between how much the knitting needles move and how long the scarf becomes. They suddenly boldly assert that they've figured out what's going on, and they proclaim that the movement of the needles *is* the scarf. This is clearly ridiculous; the movement of the needles and the knitted scarf are in no way the same thing. Despite the correlation between the needle movement and the amount of scarf knitted, there is no identity relationship between the two. You counter that the true identity relationship is between the knitted yarn and the scarf, while the movement of the needles was merely one aspect of a larger process. This is a more reasonable claim, as the scarf is indeed the same thing as knitted yarn. Similarly, if I told you that "felt experience of meaning" and "electrical activity in brain cells" were the same thing, you might reasonably disagree, as these descriptions seem to be pointing to different phenomena. An explanation of what experience is in physical terms should leave us with the impression that the same thing is actually being pointed to by both descriptions. The physical stuff of the brain is simply something different to experience, as common sense would suggest.

CONSCIOUSNESS AS COMPUTATION

Perhaps the physical properties of the brain are not necessary to explain consciousness. Some properties of things do arise from physical structure, such as the hardness of rocks or the wetness of water. Other properties are functions that do not depend on the material features of the thing in question. Consider a bicycle. It is a bicycle if you can cycle using it; it doesn't matter if it is made of wood or of metal. The function of cycling doesn't arise from this physical material but from the ability of the system to perform the function.

According to this view, known as functionalism, the mind is a functional thing of this kind and so can exist through both brains or machines; the substrate on which the function is performed doesn't matter.[7] In humans the experience of pain correlates with activity in the neocortex. Fish do not have this brain structure, but it seems plausible that they, too, experience pain. Perhaps the conscious experience of pain is related to the function of avoiding damaging events, with the physical substrate that makes this function possible being irrelevant. Perhaps it is not just pain but all of consciousness that is a function of this kind, one that can be carried out by multiple systems made of different materials. This idea is called substrate independence.

Before the advent of modern computers, functions were typically seen as being dependent on the machine implementing the function. While steam engines, cotton gins, and the telegraph could only be used to solve specific problems, humans appeared to be capable of solving problems *in general*. There were those who wondered whether it might be possible to invent a general problem-solving machine. The problems submitted to such a machine would need to be framed in a particular language that the device could understand. Mathematics has approximately four thousand years of history as a precise system for describing and manipulating patterns, making this the ideal language for such a machine. A general problem solver would have to be a machine that could manipulate numbers to perform any possible mathematical calculation. By framing real-world problems in the language of mathematics, such a machine could be used as a tool to solve any problem that had a logical solution, one that could be reached solely by analysis. The English mathematician Charles Babbage was the first to design such a machine in 1837, which he called the analytical engine. With the help of such thinkers as Ada Lovelace, Alan Turing, Claude Shannon, and John von Neumann, this machine would go on to become the modern computer.

Turing was the person who first described the substrate independence of computation. He realized that any device that could both read and write symbols to a data-storage medium, move between symbols, and control its next action could perform any possible computation. This simple process of symbol manipulation was the essence of computation, and such a device is known as a universal Turing machine. It didn't matter if it was run on a device that recorded data on magnetic tape, as Turing suggested, or on a modern computer that uses transistors or even on a set of dominoes if you arranged them the right way. It would all be computation.

The computer gives us a clear way of thinking about what a function is. A function performed by a system can be thought of as how it converts inputs to outputs. When cycling, your brain converts the sensory inputs from your eyes and the vestibular system into the output of muscle movements that keep you moving forward and not falling over. The translation of inputs into outputs is what makes this function possible. If two systems convert inputs into outputs in the same way, they perform the same function. It doesn't matter if one of the systems is a biological brain while the other is a computer; if they function in the same way, they both have minds, according to functionalism.

What is the relationship of consciousness to the brain, according to computational functionalism? The brain is seen as the hardware that runs the functional software that we call consciousness. In this view, the activity of the brain is not merely *like* a computer; it literally is computation. This raises the possibility that our technology is conscious in much the same way as we are. In a society that deeply values technological advances, the computer may be seen as a flattering analogy for our big brains. Other biological processes may be messy and off-putting, but we are reassured that our skulls house state-of-the-art supercomputers crafted by evolution to make us the dominant life-form on the planet. What's more, if your conscious mind does in fact consist of computation, then it is theoretically possible that you could upload your mind into a machine, transcending the death of the physical body. This sounds strikingly like a modern version of Descartes's immortal immaterial soul. Perhaps we haven't come so far after all.

MEANING AND SYMBOLS

Consciousness is filled with meaning, from daydreams of hopes for the future to memories of key events in one's life that can be felt to possess immense personal significance. Our mental lives can be meaningful in the sense that the contents of consciousness can feel personally relevant and important, but there is another kind of meaning that pervades our experience in every moment. This is the mundane way in which mental contents signify something other than itself. When you read these words, you understand that they mean something more than the shapes on the page. The written word *apple* signifies a real edible fruit. It has meaning that goes beyond the forms of the five letters that make up the word. Now, close your eyes and imagine a red apple. In the same way as with the written word, this conscious experience means something more than the mere appearance. There is an experience of roundness, of perhaps redness, but the whole experience signifies a food that exists outside consciousness. It is about something more than itself, a phenomenon known as intentionality. How does the apple in your mind come to be *about* real apples in the world?

Computers are capable of processing information that is meaningful to us. We have seen that computers work by turning signals from the world into a language that the computer can work with, such as binary ones and zeros, a process known as encoding. The computer can then problem solve in a wide range of domains by performing operations on these representations of the world. By analogy, the senses appear to transform signals from the world into an electrical language that the brain can use—they also appear to perform some kind of encoding. Now the brain has a "representation" of an object, some electrical signals that can stand in for the object, much like a political representative is ideally supposed to stand in for the concerns of the people they represent. The result, according to the brain-computer analogy, is that the brain works by performing operations on representations, symbols that stand in for things that actually exist in the outside world, much as computers do. The analogy is quite striking, with the digital code of ones and zeros reflected in the all-or-nothing spikes of electrical activity produced by brain cells, seemingly the brain's own kind of digital code.

The brain areas I have shown that are associated with conscious experience appear to contain representations that are experienced as conscious content. In V4 there are color signals that appear to underpin the contents of color perception. The same is true for V5 and motion, as well as for the anterior insula and a variety of interoceptive signals that arise from within the body. In these cases we appear to have electrical signals within the brain that somehow signal something outside the brain, like color in the environment or information about the state of the body. How does this occur?

We assume that our computer does not have a meaningful experience every time it symbolizes something using its transistors. Nor do we think that the words on the page of a book or the soundwaves of an audiobook are intrinsically associated with an experience when no one is reading or listening to them. What is the key difference in how the brain uses symbols, compared with the computer or the ink on the page, that makes meaning possible?

The core issue here is the question of how a symbol can come to represent something other than itself. In computers, it is us who define the meaning of a given symbol. The computer has no way to know the meaning of the information that it processes. It just slavishly performs operations on the information we provide. If you were to put numbers into a spreadsheet that correspond to the daily temperature where you live, the computer would have no idea of the meaning of this data. The human needs to be in the loop to give the computational process meaning. Without the human, computation is just a mechanistic process. This creates a major problem for the approach of computational functionalism. Who defines the meaning of the electrochemical signals shuttling around your brain? In the computer analogy, we might imagine that a symbol in the brain needs an independent observer of some kind to give brain activity its meaning, much as you define the meaning of the information you provide your computer with.

If a visual image is stored in your brain that corresponds to the visual scene in front of you, then is there another brain process that can observe this image to identify what it is about? Remember that the task here is to explain the mind. What this explanation does is simply pass the buck to another brain process that supposedly performs the function we're trying to explain. Presumably a similar process would have to take place inside this brain process, leading to an infinite regress of a brain process that observes a brain process that observes a brain process—with this approach we never close the gap between symbol and meaning. It's "brain processes" all the way down, never arriving at a final "observer." The proposal of such a mechanism is a type of informal fallacy called a homunculus (a term with an alchemical origin that you may recall from chapter 3). A homunculus is something that conveniently possesses the very capacity that we are trying to explain: in this case, a brain process that can observe other brain processes much as we observe the world around us.

This problem of how symbols or representations get their meaning was explored by philosopher John Searle with his Chinese room thought experiment.[8] The thought experiment goes like this: Imagine a person who does not understand Chinese is placed inside a room. They receive incoming Chinese symbols and are required to return Chinese symbols in response, following a detailed set of instructions in a language that they do speak. Unbeknownst to them, the incoming symbols are questions, and the outgoing symbols are answers. The person in the room does not understand Chinese and does not know that they are answering questions, yet this arrangement succeeds in allowing them to give correct responses to the questions. The aim of this thought experiment is to show that the processing rules that computers follow, those that allow them to process inputs into outputs by manipulating symbols according to rules, are not enough to generate the kinds of meaning and understanding that we find in the mind. This thought experiment suggests that the mind is not a computational function, and as a result, it would be impossible for a computer or AI system to ever become conscious. How could the logical manipulation of symbols ever produce the taste of chocolate or the smell of pine needles? A hard problem exists here between computation and experience, in the same way as it can exist between the physical world and experience. To resolve this issue, we must descend from the lofty heights of abstract computation to deal with the lowly material body.

TEN

THE EMBODIED PREDICTIVE BRAIN

Mr. Duffy lived a short distance from his body.—JAMES JOYCE[1]

THE BRAIN IN THE BODY

I ran barefoot along the muddy path next to the Thames, more exhausted than I'd ever been in my life. I had been running for twenty-three hours and was approaching the finish of a one-hundred-mile ultramarathon from London to Oxford along the river they share. The appeal of long-distance running for me largely came from the deep flow state that it produced through sheer exhaustion. On my training runs, I would find that, about an hour in, my shoulders would relax, my gait would settle into a gentle rhythmic swing, and my mind would become blissfully quiet. I would glide along canal towpaths, over hills, and through meadows with my legs and the earth seamlessly working as a whole to guide my body over the terrain. There was no separation, no friction, and no resistance—just a powerful process of surrender into this mind-altering activity. As I approached the end of my longest-ever run, my body appeared to directly feel where to place my feet on this uneven ground, my mind too exhausted to interfere in the process.

In such flow states, we typically feel that we effortlessly perceive the world around us as we engage with it. In contrast to this impression, however, contemporary neuroscience presents us with a story of perception in which our experience is mediated through representations. Perception is seen as indirect rather than direct and unadulterated. I have just shown, however, that there is the unresolved issue of how such representations or symbols could get their meaning—that is, how this indirect process might lead to conscious experiences full of significance. Perhaps there is another way of thinking about the brain that could resolve this issue.

James Gibson was a pioneering psychologist in this area who proposed an approach to the mind that he called ecological psychology. According to Gibson we do not perceive fallible representations of the world; we perceive the world directly.[2] Gibson conducted research on US Air Force pilots during World War II and observed that, when using perception to guide action, the brain is highly effective at perceiving the relationships that must be grasped in order to act successfully in the world. Pilots could apprehend how far the tarmac was from the plane when landing, as evidenced by the plane physically landing successfully. They appeared to directly grasp meaningful information about the world through the act of perception in real time.

Gibson called these properties that could be grasped directly affordances. These are the aspects of an object that permit certain behaviors. A mug handle affords holding, a floor affords walking, and a chair affords sitting, as might a fallen tree if its trunk is not too big or small. Crucially the affordance does not exist in the object itself but in the relationship between the agent and the object being engaged with. For example, one object might be sittable for a six-foot adult but not for a toddler—the affordance exists in the former case but not the latter.

This was not how Gibson's work was first taught to me as an undergraduate. By that time a divide had opened between computational and embodied approaches to the mind, and the Psychology Department at Oxford was firmly in the first camp. This is a divide that exists to this day. At that time, affordances were presented as near-mystical nonsense, a topic that I would be taught once and would never come up again. How could the brain directly perceive the sittable-ness of a chair? Was this some telepathic transmission of some ghostly affordance property physicists had yet to identify? To one who thinks in terms of separation, relational systems dynamics often appear unintelligible. This is largely why it has been necessary here to extensively explore the psychology of systems thinking in the earlier chapters of this book. Once one considers the subject as existing within the scene and not as something apart from it, we can see that there is nothing magical going on here. The affordance is a relationship between an organism and the environment it is part of, not a property of an object that is separate from the bodily subject. With Gibson's work, a new way of thinking about experience opened up. We didn't need to think of the brain as fundamentally disconnected from the environment; conscious perception could instead be understood as a process that is inherently embodied.

Embodied approaches to the mind would continue from Gibson's time up to the present day. In 1980, researchers Humberto Maturana and Francisco Varela proposed an approach to understanding cognitive aspects of the mind, such as memory and decision making, that they called enactivism.[3] Cognition is the ability to engage with the world in a meaningful way, as any intelligent or minded system does, but cognition does not necessarily require experience. An artificial intelligence could be considered a cognitive system even if we think it does not possess consciousness. In the enactivist view, the embodied organism is the relevant level at which cognitive processes should be understood, not the nervous system within the organism. They proposed that all living things perform a process they called autopoiesis (Greek for *self-creation*). Organisms contain within themselves the ability to manufacture the components from which they are made, and this process is the core of autopoiesis. The key idea is that, in addition to creating the organism, this process generates cognitive activity. In order to continually recreate oneself, one must sense, decide, and act in a meaningful way. Identifying the organism as the relevant physical basis for cognitive processing as opposed to the brain alone raises the possibility that minds could exist without brains.

Scientists need quantitative tools that can be used to study the mind so that their ideas can be put on a rigorous formal footing. The mathematical tools and principles that have been developed within computer science have been very useful for formally describing certain principles of brain function. If we do away with the computational approach, then what tools are we to use instead? There is a branch of mathematics called dynamical systems theory that studies the behavior of systems with multiple components and how these behaviors change over time. The dynamical systems approach can be used to model whole organisms as a system that interacts with an environment or to model the organism and environment as one single system. In this approach, we do away with the symbolic representations that underpin computation and look at the whole creature as an interactive process, one in direct connection with its surroundings.[4]

Some philosophers take this view so far as to say that consciousness is nothing other than a kind of movement. Enactivist philosopher Alva Noë argues that consciousness is synonymous with sensorimotor activity, and Maxine Sheets-Johnstone, a philosopher and dancer, also argues that consciousness is a kind of movement.[5] A related perspective was put forward by Robert Hanna and Michelle Maiese in their book *Embodied Minds in Action*.[6] Consider how it is that you perceive the shape of an object. You don't typically just observe it from one angle; you may move around it, pick it up, and rotate it in your hands. All perception can be thought of as an active process of this kind, like a blind person tapping their cane around their environment to actively construct a picture of it. According to these philosophers, there is no need for representation, as this interaction contains all the information necessary to perceive the object.

An issue arises with imagining objects that aren't present, however, and similarly with dreaming. Both these experiences point to some mediating step in the organism that can stand in for or represent the contents when they are not present. A synthesis of these perspectives is needed, as neither truly separate representations nor movement alone suffices to explain all aspects of consciousness. In the last few decades, an understanding of the brain has emerged that has the power to resolve the disconnect between these two extremes.

PERCEPTION AS INFERENCE

In the Islamic golden age, the medieval mathematician and physicist Ibn al-Haytham suggested that what we perceive visually is not simply the result of the information that arrives at the eye but reflects our internally generated "judgments" about what is going on in the world. In the 1800s the German physicist Hermann von Helmholtz revived this idea, proposing that conscious perception was the result of a process of "unconscious inference."[7] While we may feel as if we effortlessly perceive what is in front of our eyes when we open them, what is actually happening, according to Helmholtz, is that the brain is actively inferring what is out there in the world from impoverished sensory information.

Imagine playing hide and seek with a child. You would effortlessly be able to recognize them even if they were peering out from behind a tree and half their face was obscured. The information about their whole face is not arriving at your retina, yet your brain leverages your past experience to allow you to recognize what is really going on out there in the world. In every moment of your life, this process of predictive inference determines the contents of your conscious experience rather than the sensory information alone.

Whereas the brain-computer analogy relies on signals being encoded that can be used to reconstruct the perceived object, inference is inherently an act of going beyond the incoming evidence. When we infer something, we come to a conclusion based on available data, but that conclusion isn't present in the data itself. In the same way, the brain gets clues from the senses as to the nature of the world beyond, and it is the job of the brain to go beyond these clues to construct a detailed hypothesis of what gave rise to them. Seen in this way, the brain is inherently imaginative. Rather than simply detecting the incoming signals as a camera might, the brain is instead running something like a simulation of what events in the world most likely gave rise to the signals it receives.

In the 1990s, researchers managed to create a computer model that performed this kind of inference. When fed data, the Helmholtz machine could learn to model the hidden causes behind this information rather than the data itself.[8] This work led to the Bayesian brain hypothesis, the suggestion that the brain is fundamentally an organ of inference, a machine that is specialized for the construction of models of the world. These models should be understood as being fundamentally probabilistic or statistical; they are guesses about what is most likely going on in the world. As the millennium approached, however, one important question remained unanswered: How could such a process be implemented in the brain?

THE PREDICTIVE BRAIN

In 1999, researchers Rajesh Rao and Dana Ballard suggested that vision might be implemented in the nervous system through a process they called predictive coding.[9] This suggestion would go on to radically influence the study of the brain. I showed earlier how the primary visual cortex, V1, signals changes in the visual input rather than redundantly signaling the same information at multiple locations. In doing so, the brain manages to send information in a highly efficient manner. Given that such efficiency is synonymous with the brain consuming less energy, it makes sense that evolution would have favored brains that were wired with the appropriate connections to implement coding strategies of this kind.

This principle doesn't apply just across visual space; it applies through time, as well. If you were to watch a video of someone jumping up and down against a still background, it wouldn't be very efficient to signal the whole frame at each moment. It would save on signaling resources to assume that the background stays the same until proven otherwise. This is a fundamental principle of how neurons send information. They "habituate" or reduce their response over time if a signal remains static, only responding once a change occurs. This is why you typically are not aware of the constant touch of your clothes on your skin as you go about your day.

In addition to predicting that nothing will change in the sensory input until it does, the brain could further save on energy by having a detailed predictive model of how events are unfolding in its environment. Once you have learned to recognize how people move when jumping, you could leverage this internal knowledge to predict the pattern of movement that would be observed in the video, only signaling information when they do something other than what was predicted. By building up such models and using them to continually anticipate what will happen next, the brain minimizes the amount of information it has to signal.

There are many examples of predictions or expectations altering the contents of our conscious experience. In 2020, a video of a voice saying either, "Brainstorm," or "Green needle," went viral. In some videos, these phrases appeared on the screen as the sound repeated. Despite the sound waves remaining consistent between repetitions, individuals consciously perceived whichever word they were reading. What's more, people found it was possible to control their perception themselves by simply thinking either of these phrases. In both cases, one's expectations of what words were being spoken shaped what was experienced. The contents of our consciousness are clearly grounded in the expectations or predictions formed by the brain, as well as the sensory evidence that it receives.

THE CONTROLLED HALLUCINATION

The predictive-processing approach validates Kant's distinction between the phenomenal world of experience and the noumenal world beyond.[10] When we perceive the world around us, we are not coming into direct contact with reality but are inferring what it might be like. We experience the image of the world that we have constructed, not the world itself.

What we experience depends both on our internal models and on the sensory signals our brains receive. You can experience this fact for yourself. The Kanizsa triangle illusion consists of what appears to be two white triangles with a partially obscured black circle placed underneath each corner of the topmost triangle (see figure 10.1). Upon closer inspection one notices that this apparent triangle is an illusion; there is no such shape present. Instead, there are only three Pac-Man-shaped figures and a collection of lines that merely imply the form of a white triangle. As a result of your brain developing in a world in which triangles and circles exist and are

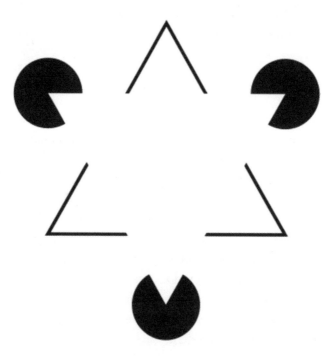

Figure 10.1.   The Kanizsa triangle illusion produces the impression of a white triangle that is not actually present in the image. *Wikimedia Commons.*

capable of obscuring each other, your brain infers that most likely two triangles and three circles are present. The triangles are not present in the sensory data, yet it is what we perceive because our brains infer their existence.
We construct a belief, model, or hypothesis about the existence of the white
triangle, and it is this guess about what is really going on in the world that
shapes our conscious perception.

We are born with some hard-wired expectations about the structure of
the world. For example, your two eyes can be interpreted as an expectation that you would be born into a three-dimensional world where depth
exists, as the placement of your eyes on your head is used to extract such
information. Most of our predictive models, however, are acquired through
our cognitive development. In artificial neural networks that function by
predictive-processing principles, the system typically is first trained to recognize objects and, in doing so, to build up an internal recognition model.
When run in reverse, this recognition model becomes a generative model
that can predict what input it should be receiving. This kind of logic underlies generative AI algorithms that can produce images from text prompts,
as well as the phenomenon of dreaming.

HIERARCHICAL PREDICTIONS

We do not infer the whole picture of what is going on around us in one
single leap of intuition. The brain is thought to be arranged hierarchically,
with many specific guesses about what is going on in the world being nested
within big-picture guesses.[11] In the case of the illusory white triangle, the
big-picture belief that there is a triangle initially overrides the more nar-
rowly focused beliefs that there is no edge where there should be. If you
focus your attention on one such illusory edge, the experience of the triangle
may temporarily disappear, as you bias your brain into putting more weight
on these lower-level specific beliefs. Despite such contradictory beliefs being
found in the same hierarchy in your brain in this example, they typically do
not coexist peacefully, and in most cases, one wins out and determines what
it is that you perceive in any moment.

In this view of how perception is implemented, we start out not with
sensory input but with the brain's big-picture guess about what is going on.
In the hierarchical organization of the brain, these most general hypotheses

about the world exist at the top of the hierarchy, away from the sensory input, with more specific beliefs existing in the lower layers of the hierarchy. The brain's big-picture guess is then used to make predictions about what signals will be received by the senses in the next moment. If the expected signals arise, there is no need to waste energy signaling anything further, and the system can continue on its course. If something unexpected does happen, then a prediction error occurs, and this error signal is passed up the hierarchy so it can be used to update the beliefs that the brain holds. Prediction errors flow in a bottom-up direction from the senses toward the top of the brain's hierarchy, but the content of experience is largely determined in a top-down manner, originating in the brain's high-level beliefs.

HOW TO INFER

We perceive what we infer to exist, not necessarily what really exists. Mathematicians have long understood how to optimally make inferences based on evidence. Thomas Bayes was a philosopher and Presbyterian minister who developed the theorem that bears his name. Bayes' theorem captures how beliefs can be optimally updated in light of new information. You start with a particular belief in whether it has rained recently, for example. If you see a wet road, then you may want to update that belief in whether it has rained. According to Bayes' theorem, in order to update one's beliefs in the optimal way, one must consider not only the new evidence but also the relationship between the probability of rain and the probability of the road being wet if it had indeed rained, as well as the overall probability of observing a wet road.

Bayesian inference starts with a prior-probability distribution, which captures the relationship between evidence and their causes prior to any new evidence being received. If you live in southern California, then the likelihood of a wet road being caused by rain in August is low, whereas in London it's a different story. The prior-probability distribution, or prior for short, is different for each location. As new data becomes available, you update your beliefs to obtain a posterior-probability distribution, which captures your new belief at the end of this process of inference.

Reasoning in line with Bayesian principles means choosing the best explanation for a piece of evidence, weighted by how likely it is as an explanation in general. One study found that people were more likely to guess that someone described as being "very shy and withdrawn . . . helpful . . . meek and tidy" with a "need for order and structure, and a passion for detail" was a librarian rather than a farmer.[12] This may seem like a good hypothesis based on this piece of evidence, but this reasoning is not Bayesian, as it does not weigh the hypothesis by how likely it is to be true independently of the evidence. In the United States, there are approximately fifteen times as many farmers as there are librarians. Once one factors this information into one's calculation, it becomes far more likely that this individual is simply a shy farmer rather than a stereotypical librarian. By incorporating more prior knowledge, our hypotheses or beliefs about what is going on in the world can be made far more accurate. If we continue to receive evidence over time and we use this evidence to refine and update our beliefs in this way, then we can approach the truth of the situation as quickly and effectively as possible.

It has been suggested that the brain uses a Bayesian algorithm of this kind to update its models of the world in light of evidence from the senses. This idea is known as the Bayesian brain hypothesis. What form might Bayesian beliefs take in the brain? Rather than consisting of a single probability value, the brain typically deals with distributions of probabilities. For basic sensory input, such as a vertically oriented line, there might be a bell-curve-shaped distribution centered around the belief that it is vertical and tapering off toward horizontal, capturing the hypothesis that vertical is most likely but that there is some uncertainty. If the bell curve is wide, resulting in there being some probability above zero that the line could in fact be horizontal, then this prior belief would be quite uncertain. Technically, this is referred to as having low precision. If the bell curve was very tightly centered on vertical, then the prior would be very precise—it would have high precision. The precision captures our certainty about a belief and factors into how our beliefs are updated. If our prior belief has low precision, then we are not confident in its accuracy, so it should be updated to a greater extent in light of new evidence, compared to a belief we have high confidence in. If I'm certain that I'm looking at a cow, then it's going to take a lot of strong evidence for you to convince me that it is actually a cat. If I'm uncertain of what I'm looking at because it's a foggy day, then I should rely more on the evidence you offer me rather than sticking with my uncertain guess.

The brain's models of the world exist as probability distributions of what could be occurring. How might these probability distributions exist physically in the brain? One proposal is that they could be signaled via the frequency of electrochemical pulses, or spikes, produced across a population of neurons. The frequency of spikes that a neuron produces is known as its firing rate. Another way that they could be represented in the brain is through changes in the strength of connections between neurons, which can be thought of as representing the weight of different hypotheses. Additionally, these hypotheses can be encoded in temporal dynamics, such as the coherence or phase of neural activity. We do not currently have a complete picture of how inference is performed by the brain, as the precise mechanisms of predictive processing in the nervous system are an area of active research.

This would be a great algorithm for the brain to use, except that calculating all the possible causes of the evidence we receive is practically impossible in the real world. In the earlier example of observing a wet road in southern California versus London, we need to know the overall probability of observing a wet road due to any cause. This involves having access to the probability of rain causing a wet road, of sprinklers being responsible, of a car wash fundraiser in the street, and any other number of possible causes. When the brain is presented with a signal from the outside world, there is no way for it to know of the possible causes for that signal. Imagine trying to calculate the likelihood of whether a vertical line in your visual field is present due to looking at the edge of a building, a kitchen cupboard, a table leg, a block of wood, and on and on. The brain simply can't achieve this feat. Fortunately for us, there is a practical workaround, a way of approximating Bayesian inference without having to perform this impossible calculation.

THE FREE-ENERGY PRINCIPLE

Rather than performing perfectly accurate Bayesian inference, the brain appears to be able to approximate this process via an elegant strategy. This strategy consists of starting with any guess at all about what the cause of the evidence is and then updating this guess based on how surprising the evidence is in light of it. It is effectively the strategy we instinctively employ in a game of twenty questions. Imagine if you were to hold out an object in front of me while my eyes were closed, and I had to determine what it was. I start with a random guess that it's an apple. You then tell me that it is edible. This is entirely consistent with my guess that it is an apple, so it is not a surprising piece of information at all if I am correct. You tell me it is roughly spherical. Same story: I stick with my guess that it is an apple. You then tell me that it is brown. Red or green would not have been surprising, but this is a slightly more surprising piece of evidence if it is an apple. It's still possible it's an apple, but this evidence doesn't fit as neatly as it being red or green. As a result, I slightly shift my guess away from it being an apple to it being a potato. Finally, you tell me that it can be made into French fries. This is highly inconsistent with it being an apple so I firmly shift to the confident guess that it is a potato. By starting with a random belief and updating it in line with how surprising the evidence would be if my belief were true, I manage to converge on the correct belief as efficiently as possible with no need to know all the possible causes behind the evidence I'm receiving.

The proposal that the brain follows a surprise-minimizing strategy of this kind to implement approximate Bayesian inference is known as the free-energy principle (FEP).[13] Professor Karl Friston, the most highly cited neuroscientist in the world, is the architect of the FEP. The FEP has been used to explain why what we perceive depends on both the sensory evidence we receive and our prior expectations of what is giving rise to that sensory evidence. Perception gets going with the process of constructing guesses about the world and generating expectations and then acts to minimize how incorrect these expectations are going forward.

FROM PERCEPTION TO ACTION

We've seen how perception can be understood as being implemented through predictive processing in the brain. If you were walking in the savanna and a lion were to suddenly appear in front of you, your brain could efficiently allocate processing resources to the new, unexpected signals it receives, allowing you to perceive the lion. If this were all your brain was capable of, however, you wouldn't last long. Your ancestors who survived to pass on their genes also had to be able to act in the world, to run away from the lion in this case. Did a whole separate brain process have to evolve in order to execute behaviors of this kind, or could the very same prediction machinery of the brain be leveraged to achieve action, as well as perception?

When you reach for an object, many different muscles in your arm have to contract and relax in precisely the right way to get your hand into the correct position. One way to achieve this would be for your brain to laboriously calculate all the different necessary lengths of the muscles and then send signals to each one, indicating how much they should relax or contract. Another option would be to simulate the end result, move the muscles a small amount, and check whether you are getting closer to the simulated goal state. By performing this process continually, your hand could move directly toward its destination in the most efficient manner possible. It appears that this latter option is what the brain does.

Voluntary movement, according to this view, is initiated by the brain predicting the sensory signals that it would receive if the action were successfully performed—for example, the sensation of fur on your hand if your intention was to pet a dog. The brain then initiates actions to minimize the prediction errors that arise due to the difference between the current state of the body and the state it would be in if the action were successfully completed. Using this strategy, your hand could be directly guided to its correct final state, with all the different muscles smoothly being guided into position, as if by magic, pulled into place by the hidden puppeteer that is your brain. How much easier it would be to win on a claw-machine game at a fairground with this strategy, if you could simply instruct it to achieve the directed goal state of picking up your chosen stuffed toy. The machine would then automatically move in all three dimensions at once to execute your wish rather than you clumsily shifting in all three dimensions independently, only stopping once it had achieved the desired goal.

This incorporation of voluntary movement into the predictive-processing framework is known as active inference, which can be contrasted with the process of passive inference that underlies perception.[14] Your brain continually shifts between these modes of active and passive inference. When we are not initiating voluntary movements, we are open to using the evidence received by our senses to update our model of the world. When we act, we fix our model of the world and treat our beliefs as commands to be brought into reality rather than interpretations of what is currently occurring. You get out of bed in the morning because the brain constructs a high-level belief that this is what is happening. The surprise that arises in the system from the sensory feedback that you are still in bed is then minimized by you moving your muscles until you are standing up. The belief that something else is happening in this moment is a self-fulfilling prophecy that brings that situation into existence.

According to the FEP, action and perception are not separate functions performed by the brain but are two sides of an embodied process. It is not the brain alone that constructs beliefs about the world. Instead, these beliefs exist in an embodied feedback loop that involves both the body and the world. In this picture, the external state of the world, the sensory information received by the brain, the actions performed by the body, and the internal states of the brain all function together as a seamless whole. The state of the world and the previous action taken by the body determine the sensory signals that will be received, while these signals and the previous internal state of the system determine the next internal state of the system, and these internal states and the sensory signals received determine the action that will be taken, which affects the sensory signals that will be received, and the feedback loop continues. Through this embodied behavior, beliefs about the nature of the world beyond the organism are sculpted and honed through the process of surprise minimization, in which beliefs are updated in line with how surprising new evidence is. We should not think of these beliefs as existing solely in the brain, however, but in this entire loop, from internal states to action to world to sensation to internal states. Holding beliefs is an embodied process.

This approach sidesteps the issue in the previous chapter of how representations can come to correspond to the world outside the organism. In the encoding approach, the representations in the brain are re-presentations of the objects in the world that gave rise to the incoming sensory signals. With this picture, a mystery remains about how the correspondence emerges between the thing in the world and the representation of it in the head. There is a structural chain of causality that links the object to the representation via signaling in the nervous system, but Searle's Chinese room thought experiment shows that this kind of syntactical pattern never gives rise to semantic meaning. In the predictive-processing approach, however, representations can be understood as hypotheses that are sculpted by the selection algorithm of the FEP through the feedback provided by prediction errors to correspond sufficiently to the structure of the world beyond in order to be of use. A correspondence between the internal picture of the world and the actual world emerges as a result of this process of inference. Here we have an explanation of how a certain representation-like phenomenon can exist in the brain and can come to refer to things other than themselves.

PREDICTIVE PROCESSING
AND CONSCIOUSNESS CONFUSION

There is much excitement and debate about the potential of the power of the predictive-processing perspective to contribute to our understanding of consciousness. Multiple theories of consciousness have been proposed that link experience to processing in the brain of this kind. These include Andy Clark's generative entanglement, David Rudrauf's projective consciousness model (PCM), Adam Safron's integrated world modeling theory (IWMT), and Anil Seth's beast machine theory, among others.[15] While these theories tend of focus on the cortex and how it produces our perception of the world around us and ourselves within it, another approach has been proposed by Mark Solms, which links consciousness to the brainstem and its role in interoception and affect (i.e., the detection of signals that arise from within the body and the subjective experience of emotion).[16] This split between the cortex and brainstem camps when it comes to the brain basis of consciousness is explained in an earlier chapter, and the debate continues when it comes to viewing consciousness from a predictive-processing perspective.

With the proliferation of theories regarding how predictive processing in the brain could underpin consciousness, researcher Wanja Wiese has suggested that the field should start focusing on synthesizing these proposals into a minimal unifying model, a theory that captures whatever the common core is of all these competing theories.[17] Such a minimal unifying model has been proposed in the form of the inner screen model of consciousness. Based on the work of Karl Friston and Chris Fields, researcher Maxwell Ramstead teamed up with these thinkers, as well as with Adam Safron, Mark Solms, and others, to develop this attempt at a minimal unifying model. Their proposal is that, at their core, these varying models can be boiled down to the idea that the topmost layer of the predictive-processing hierarchy functions as a kind of self, a naturalized version of Descartes's homunculus.[18]

No matter the specific mechanism proposed, however, we come up against the hard problem yet again. How can a description of a brain process be used to explain a qualitative experience like the taste of orange juice? The information-processing dynamics can be run on a computer, as with the Helmholtz machine, so what is it about the brain that gives rise to phenomenal experiences while the computer presumably does not? Unless you think the computer does have experiences, in which case we are back to computational functionalism. At this point it may feel as if we should simply lock ourselves up inside Searle's Chinese room and throw away the key.

Anil Seth has argued that we should not be focusing on resolving the hard problem but on what he calls the "real problem of consciousness."[19] This refers to the day-in-day-out attempt of science to slowly but surely shed light on whatever phenomenon it chooses to study. The hope is that research in the end will dispel any desire to ponder the hard problem. Once we know enough about the brain, we'll simply find there is no mystery to be solved. Along with Jacob Hohwy, Seth has proposed an ambitious research plan to do just this.[20] But this belief that science can tackle consciousness in the same way as the other publicly observable phenomena that it studies is exactly what led David Chalmers to propose the hard problem in the first place and, in doing so, to point out that this logic does not hold when it comes to understanding experience. We cannot ignore that fact that our current worldview has proven incapable of explaining consciousness.

ORGANIZATIONAL FUNCTIONALISM

The world, somebody wrote, is the place we prove real by dying in it.
—SALMAN RUSHDIE[21]

The embodied and predictive perspectives explored here see the behaving body as providing an important contribution to the process of conscious perception in brains. Where does this perspective fit into the division explored in the previous chapter between consciousness being viewed as a substrate-independent function versus it being grounded in the physical structure of the brain? It is common for the predictive-processing perspective to be interpreted in terms of computational functionalism, but I believe it holds the potential to reconcile these differing perspectives. To understand how, we must return to the distinction between form and function.

Consider a dining table. Its material form gives it certain properties. For example, if it is made of metal, it will have the property of being hard and resistant to bending, but if it is made of a thin plastic, it may not have this property. Its property of being a table, however, exists not because of the material it is made of but its function. It is a dining table if you can sit at it and eat your dinner off it. A table could be made of wood, metal, plastic—it doesn't matter as long as it functions as a table. My grandmother's first dining table was an ironing board, yet it was possible for it to also *be* a table because it performed that function.

This is a very neat distinction between form and function, but consider if they are truly as separate as I have just made them seem. While the function does not arise out of the material properties of the table alone, the object can only function as a table if the overall form has certain characteristics. It must be a few feet high—no more no less—and must be rigid enough to rest a plate of food on. A wet carboard box cannot be a table, nor can something eight feet high; their forms prevent them from performing that particular function. Far from being distinct, the overall form actually dictates function.

This perspective is a kind of organizational functionalism, as the core idea is that the overall organization of a system determines its function. In the case of the beliefs constructed by the brain, it is the fact that they are embodied in a feedback loop with the environment that gives them their function; they do not exist as an ephemeral computation in the nervous system but as a relational process between embodied creature and environment. Here, mental functions exist as the organized interactions between the organism and the wider world; function and organization are the same thing. We can surmise that the fish and the human both feel pain when harmed despite having different nervous systems because the function of pain is to avoid harm, and the forms of both organisms embody this function of avoiding bodily harm. If organizational functionalism is correct, then we must look to the organization of the whole organism in order to understand consciousness and not to abstract computation or the brain alone.

This shift from the brain and toward the whole living organism can also explain how it is that the content of consciousness takes on a further kind of meaning beyond simply standing in for things other than themselves. In addition to this capacity to represent things, the contents of consciousness also come to take on personal significance of great meaning for the organism having the experience. It's one thing to represent a neutral object outside oneself but another for it to carry this kind of significance. For example, imagine the importance that food is felt to have from the position of a hungry person. The significance here arises in light of the individual's drive to stay alive, and it is this process that ultimately endows all conscious contents with varying levels of significance. Living systems persist over time, and everything they do is sculpted by this dynamic of persistence. To persist over time in this way is to act out a preference for survival, which is the same as preferring not to die. It is this fundamental preference for life over death that gives the contents of experience significance—meaning only exists in light of death.

This idea has been explored by multiple thinkers, such as Terrence Deacon in *Incomplete Nature* and Mark Bickhard in his interactivist model of representation.[22] It is a key part of the study of biosemiotics, how it is that living things imbue the world around them with meaning. It is also captured in the earlier theory of autopoiesis and in the work of advocates of this perspective, such as Evan Thompson. Relatedly, Artemy Kolchinsky and David Wolpert, researchers at the Santa Fe Institute, have proposed a measure of semantic or meaningful information for a system based on the extent to which that information can be leveraged to support that system's survival.[23]

Again and again, the science is pointing toward the living body as crucial for multiple aspects of consciousness. In order to understand precisely what it is about biological systems that enables them to be conscious, we must attempt to understand the fundamental nature of life itself.

ELEVEN

WHAT IS LIFE?

From all we have learnt about the structure of living matter, we must be prepared to find it working in a manner that cannot be reduced to the ordinary laws of physics.—ERWIN SCHRÖDINGER[1]

There is a video on the internet called "Single-Celled Organism Dies" that, at the time of writing, has been seen more than 10 million times. Posted on the YouTube channel Jam's Germs, the organism in the video is of the genus *Blepharisma*. In the video, the single-celled creature is moving around using small limbs, when its internal contents begin to leak out of one side of its membrane. It manages to stem the loss and keeps moving, but then another leak erupts on the other side of the membrane. It appears to struggle on valiantly, almost looking like it is going to survive. It swims away from its spilled insides, managing to hold itself together, the seconds ticking by as the viewer waits for our hero to meet its inevitable fate, predestined by the video's title. Suddenly the membrane bursts where its limbs are, the whole cell wall collapses, and an instant later, there is no longer life where there once was. One moment there was a complex creature navigating in the world and trying to stay alive; the next, only a soup of chemicals with no inside or outside.

The approximately 50,000 comments on the video include deeply empathic sentiments, such as "It's painful to witness the struggle before it dies," and "I am struck with a tremendous amount of compassion for this simple little organism." The most common sentiment is one of surprise at the outpouring of empathy for such a small creature. Another writes, "I'm confident this thing is essentially a biological robot so there's no consciousness being snuffed out, but it is an interesting question where that line is drawn. A nematode? A mite? A frog?" A modern-day Descartes responded with "actually thinking of it as a robot makes me feel better."

VITALISM

The human mind evolved to believe in the gods. It did not evolve to believe in biology.—E. O. WILSON[2]

We are often given a picture of the brain as the command center of the body, the CEO that sends orders for the body to obey. While it is true that the brain sends signals that control the operation of the body, it is important to remember that this all occurs in service of the life process. The brain evolved to serve the living body; it is a powerful servant rather than the master. To fully understand the brain, we must first understand life.

Living things look very different from nonliving things, and it is clear that these two phenomena are quite distinct. While we are alive, the body exists as a vital, dynamic, organized form that metabolizes food into energy, radiates heat, and can make change in the world. With death this all goes away. For a long time, it was believed that this difference must be due to some special vital substance or force that exists in living things, an idea known as vitalism. I've shown that scientific explanations often begin with a naive intuition that the phenomenon in question can be explained by appealing to the existence of a special substance, only for this way of thinking to be replaced with a process-based understanding. The idea of a vital substance was an early attempt to account for the reasonable intuition that living things are dramatically different from nonliving things.

Some vitalist thinkers suggested that life was due not to a special substance as such but to a force not accounted for in the existing laws of physics and chemistry. Spinoza used the Latin term *contaus*, meaning *striving, effort*, or *endeavor*, to refer to a fundamental imperative that is responsible for both the motion of nonliving objects and the self-persistence of living things.[3] Schopenhauer took up a similar idea, referring to the "will to live."[4] Continuing this tradition, Henri Bergson would develop his famous concept of the *elan vital*, or vital impulse, in the early twentieth century.[5]

If we interpret these "forces" as a dynamic or mode of organization, then we are onto a more reasonable proposal than that of a magical substance or force. There is clearly something different about the behavior of living versus nonliving things. If you randomly rearrange all my atoms and molecules to disrupt my overall organization, I will no longer do any of the interesting things we associate with systems that are alive. What is it about the organization of an organism that makes it a living thing? By studying life as a process, we have the chance to account for the key difference between life and nonlife without appealing to some magical life-giving substance.

FROM PHYSICS TO LIFE

Life is often defined in a descriptive manner. We generally observe living things as possessing the ability to regulate their internal environment (homeostasis), maintaining organization, metabolizing food into usable energy, exhibiting growth, displaying adaptability, responding to stimuli, and having the capacity for reproduction. There are always exceptions to descriptive definitions of this kind, however, such as mules, which cannot reproduce and are still considered to be alive.

To understand what life is at a more fundamental level than superficial descriptions, we must look to the intersection of physics and biology. From the perspective of complex biological systems like ourselves, single-celled organisms look comparatively simple. From the perspective of physics, though, they can appear to be something approaching a miracle. The physicist Erwin Schrödinger, of Schrödinger's-cat fame, turned his attention to the physics of life in a series of lectures he gave while working at Trinity College, Dublin. The material from these lectures would be published as his seminal text, *What Is Life?*[6]

Schrödinger put forward a conception of life as an island of self-maintaining order. Multiple terms have been used by different scientific communities to refer to this crucial characteristic of living things. In thermodynamics it is *staying far from equilibrium*, in cybernetics it is *persistence*, and in biology it is *homeostasis*. In order to understand what Schrödinger intended when characterizing life in terms of order, we must consider the thermodynamic perspective. Thermodynamics began as the study of how heat moves so that it could be used in the control of technological systems during the Industrial Revolution. The paradigmatic example is the use of hot water in the form of steam to power steam engines. The principles of thermodynamics turned out not just to apply to technology, however, but also to capture fundamental principles of how energy operates in our universe.

Time plays an important role in thermodynamics. Why is it that time flows in one direction and not the other? Hot water cools but never spontaneously heats up. Eggs break into pieces, but the pieces never suddenly reassemble into a whole, unbroken egg. The consistent directions of these processes are due to the second law of thermodynamics. The first law states that energy can never be created nor destroyed but instead only changes its form. The second law states that, in a system that doesn't receive energy into it and can therefore be considered closed, these changes in form always move in the direction of greater entropy. Entropy can be thought of as disorder or, more accurately, how many arrangements of the parts of a system there are that all produce the same organization of the whole. Milky coffee is highly entropic because you can shuffle the molecules that make it up in countless ways, and it will still look like boring old milky coffee.

What are order and disorder? When something is orderly, it is arranged in a particular way. By definition, orderly states are outnumbered by disorderly states because something that is disorderly doesn't have to be a particular way. A collection of grains of sand is considered to be a sandcastle when they are organized in a specific arrangement to form walls and turrets, but a randomly organized pile is just a pile of sand. Let the wind blow on the sandcastle and the waves crash into it, and these random motions will most likely result in a sloppy, disordered sand pile. There is nothing in the laws of physics that prevents a sandcastle from being spontaneously formed by these forces; it is just incredibly unlikely because the grains would have to end up being organized in a very particular way.

Life bucks the trend toward disorder. Living things manage to keep their component parts organized and orderly over time. They are examples of stable order for the time that they are alive. A living thing is like a self-sustaining sandcastle that manages to keep itself together, despite the potentially destructive forces of the world around it. In general, our skin closes up and heals after damage, we effortlessly maintain our temperature and other bodily states in precisely the right range for life (the phenomenon of homeostasis), and we attempt to ensure we have food and shelter to keep ourselves surviving day in and day out. Does this order-maintaining behavior of life violate the claim of the second law that all closed systems become disordered over time, moving toward greater entropy? No. The second law states that only *closed* systems that receive no energy from the outside tend toward disorder. Living things are open systems that take in energy and, as a result, can maintain their order, as long as the overall disorder of the universe as a whole increases.

Our planet is drenched in free energy from the sun, energy that is available to do work. While the sun itself may look like a highly disordered riot of energy, the radiation we receive from it is actually highly ordered in a thermodynamic sense. This energy is called free energy as it is available to do work. Schrödinger called the kind of order we receive from the sun in the form of low-entropy sunlight negentropy, a term that has since fallen out of favor. He suggested that life can exist because it borrows its order from the sun, allowing us to persist for a time amid the trend toward decay.

Change occurs in our universe where energy gradients exist. The energy of the sun radiates outward because the sun has more energy than its surroundings. It is as part of an energy dissipation process of this kind that interesting forms like life can arise. In classical thermodynamics, the energy gradient is eliminated by a straightforward mingling process, one that produces a homogenous result, known as a state of equilibrium. Think of what happens when cream is added to coffee. There is a strong gradient in terms of where there is and isn't cream at first. Over time, the coffee and cream mingle until the result is a homogenous blend of coffee and cream. At this point, it has reached equilibrium.

At equilibrium, nothing much interesting happens. When a system is in a state of disequilibrium, complex kinds of order can emerge. The study of such systems is known as nonequilibrium thermodynamics. In the coffee example, complex tendril patterns arise as the two liquids mix, but no stable forms are produced. Interesting forms can be produced by the dissipation of energy gradients in some special cases. When a pan of fluid is evenly heated, it is possible for structures called Rayleigh–Bénard convection cells to emerge. These are stable hexagonal patterns that spontaneously arise in order to dissipate the heat optimally. Another example of emergent order of this kind is the whirlpool that appears when draining a bath.

The Nobel Prize–winning chemist Ilya Prigogine coined the term *dissipative structures* for these orderly forms that emerge so as to eliminate an energy gradient most effectively.[7] A living thing is a kind of dissipative structure. The universe permits you to exist because you do it the favor of breaking down the natural products we call food more effectively than if they simply decayed without us chewing and digesting them. We maintain our local order at the cost of contributing to the overall trend toward disorder in the universe. If you want to make the inside of your refrigerator colder, thereby locally decreasing its entropy, the machine will pay for it by using up more energy to do the work of cooling and by radiating more heat into your kitchen. This is why running your refrigerator with the door open on a hot day will not cool the room down; it will actually add to the heat. There is no cheating the overall tendency toward greater entropy.

When hot and cold air mix, a tornado may temporarily form until equilibrium is reached. The tornado is a dissipative structure, one that exhibits a relatively stable form for the time the gradient of temperature exists. More than this is going on when it comes to the physics of life, however. Crucially, living things adapt and evolve over time toward orderly arrangements that manage to harvest energy, energy that can be used to do the work of maintaining these orderly arrangements.

The difference between the tornado and a growing plant becomes apparent if you imagine watching a movie of the tornado forward and in reverse. If the tornado was in an empty landscape, it would be difficult to tell the difference between these two versions of the same movie. In the case of the growing plant, the reverse movie of it shrinking back into the ground would be easily distinguishable from the phenomenon of growth. Living things evolve over time toward orderly forms that are unusually good at harvesting energy, compared to random arrangements of the same physical parts that make them up. The asymmetry in time of such processes, their being "time-irreversible," turns out to be crucial to explaining this basic aspect of life.

There are many more ways to have the parts that make up a plant lying around in a disordered fashion, compared to the very specific arrangement that characterizes the functioning plant. Can this phenomenon be incorporated into our understanding of physics? The second law has recently been extended to do just this. In one experiment, researchers put forty beads made of the electrically conductive material chromium into a dish filled with oil.[8] They then provided an electrical energy source at the center of the dish. The beads spontaneously arranged themselves into "dynamic tree-structures" that moved in a wormlike manner and "healed" themselves when they were disturbed. Compared to beads moving around randomly, the evolution toward this specific structure is highly time-irreversible; the forward and backward movies would look very different. Physicist Gavin Crooks showed that the asymmetry of such processes is always linked to a release of heat: Things taking on a specific orderly structure through time must be paid for with an overall increase in entropy.[9] Fellow physicist Jeremy England has generalized this insight to account for the phenomena of self-assembly and energy harvesting that we observe in living things.[10] He calls the process that accounts for these phenomena dissipative adaption.

It might appear very unlikely for such systems to undergo the precise series of steps necessary for it to end up capturing and using up an energy source. That would be like the components of a solar panel being left in the sun and then somehow sorting themselves into the correct arrangement. If we instead consider molecular systems in a heat bath, however, the sequential building of such arrangements becomes vastly more probable due to the relationship between energy dissipation and time irreversibility. Arranging the components in a precise energy-using arrangement takes work, and this work uses up some of the energy that is available. As energy is continually used up to create order, the system becomes ratcheted further and further into an organized state. This is the kind of time-irreversible process I explored earlier with the phenomenon of biological growth. As a result, energy dissipation favors the emergence of order. Where an energy gradient exists to be dissipated, orderly arrangements emerge as a result of the energy flow through the system, arrangements that would otherwise seem vastly improbable. Overall energy dissipation leads to the emergence of energy-channeling orderly structures in such circumstances.

For a long time, the emergence of life on earth was assumed to be the result of a vastly improbable fluke occurrence. In contrast, the phenomenon of dissipative adaptation makes the emergence of life appear as an inevitable continuation of the energetic dynamics of the earth. This would explain why life seemingly emerged as soon as it was possible for it to, after the earth cooled. We can think of the driving force behind this phenomenon as the dissipation of gradients of order that is captured by the second law. Imagine a flooded bathroom above the ground floor of a building. As the pressure from the water's weight increases, it will naturally find the path of least resistance to release the potential energy produced by the effects of gravity, inevitably leaking down into the rooms below. In the same way that water will find the cracks in the ceiling to dissipate this gradient, the energetics of the earth found certain "cracks," in the forms of the complex chemistry that became life, through which it could most effectively dissipate the geothermal energy gradient. As Jeremy England puts it, "You start with a random clump of atoms, and if you shine light on it for long enough, it should not be so surprising that you get a plant."[11]

With this understanding we move a step toward life, but we are not there yet. This kind of self-organization is not enough to capture the full complexity of living things, not least because these processes do not replicate themselves. While Schrödinger argued that all life actively maintains its own existence as an island of order, he also proposed that this ability to replicate is another crucial feature of living things. In 1944, he speculated that cells might contain what he called an aperiodic crystal, an irregular arrangement of chemical bonds, that would be capable of storing the information necessary for the replication of the cell to occur. This speculation was astonishingly prescient, with it leading researchers Rosalind Franklin, James Watson, and Francis Crick to discover the double-helix structure of DNA. DNA was the information-storing aperiodic crystal that Schrödinger had envisioned.

In keeping with the reductionist bias of modern science, the discovery of DNA led to genetics taking center stage in the following decades, with the importance of the orderly holistic organization of life and the cell fading into the background. Debates arose in origin-of-life research about whether the process of replication or the process of metabolism, by which cells extract energy from food to keep themselves orderly, arose first. Discovering the specific details of what happened for life to begin 3.7 billion years ago is a tricky business, and we have yet to resolve this question one way or the other. What we do know is that at some point both these processes came together, giving birth to life as we know it.

THE ORIGIN OF LIFE

What specific circumstances gave rise to the carbon-based life-forms that we find on earth today? Since the times of ancient Greece up until the seventeenth century, the common assumption was that many life-forms sprung spontaneously from inanimate matter, an imagined phenomenon termed spontaneous generation. Maggots were supposed to arise from rotting meat, while frogs were thought to grow out of mud. The physician Jan Baptist van Helmont went so far as to document a recipe for creating mice, involving placing a piece of dirty cloth in some wheat for three weeks, while he suggested one could create scorpions by placing some basil between two bricks and leaving them in sunlight.[12] Thankfully, our understanding of how inanimate matter became animate has come a long way in the last few hundred years, and I do not present you with such recipes here.

Let us zoom in on the specific properties of the simplest form of life around today, a single-celled organism, in order to see what ordered structures had to come into existence for terrestrial life to begin. Single-celled organisms are contained within a plasma membrane that separates them from their environment. They possess a metabolism, the ability to take in energy and use it to make their own component parts. They also possess the ability to replicate. These characteristics of the cell all seem to depend on each other for their existence, creating a three-part chicken-and-egg problem when it comes to how life got started. Researchers disagree about which of these processes emerged first, with multiple proposals having been put forward.

Some researchers have proposed that the ability to replicate was the first to come into existence. According to this view, life began in a situation that has come to be called RNA world.[13] RNA is related to DNA and is found in all living things. It is essential for almost all biological functions, which it achieves through either the building of proteins that perform the function or by performing the function itself. According to the RNA-world hypothesis, chemical processes first produced RNA molecules, and these strands of RNA then began spontaneously replicating, with mutations arising, allowing for Darwinian selection to occur in these complex molecules. The idea that strands of RNA could evolve over time in this way without a metabolism to support them, however, is associated with many unresolved questions, such as why they were not destroyed by the water they existed in and why they developed into life rather than a tar-like organic sludge, a result that would appear more likely.

Another possibility is that the membrane of the cell, rather than replication, came first. According to the lipid-world hypothesis, life began with the fatty molecules that naturally ball up to create spherical bubbles, bubbles within which the chemistry of life could flourish as a result of being insulated from the outside world.[14] When they get large enough, such fatty bubbles naturally pinch off a portion of themselves, splitting in two and potentially laying the foundation for the process of cell division. If RNA molecules became trapped inside these bubbles, then they may have been sufficiently insulated from the destructive forces of the outside world for selection to act on them, allowing life to get up and running. From a thermodynamic perspective, however, this "just so" story skirts around arguably the most important aspect of living things: their metabolism. We must explain how islands of order arose that could harness energy to keep themselves together over time if we are to understand the origins of life.

Might metabolism have been the first aspect of life to come into existence? Life is essentially complex chemistry that arose out of simple chemistry. One chemical process in particular has been proposed by complexity theorist Stuart Kauffman to lie at the origins of life and its metabolism: the autocatalytic set.[15] A catalyst is something that increases the rate of a chemical reaction. An autocatalytic set is a set of chemical reactions that catalyze themselves, creating a self-amplifying feedback loop, a phenomenon termed catalytic closure. Research has found that these autocatalytic sets are capable of undergoing evolution. Life may have begun with a simple chemical feedback loop of this kind that then evolved into the complex metabolic networks that we find in cells today.

I am still talking here about general principles. What do we know about the specifics of how life started here on earth? In the first half of the twentieth century, the idea that the structures that make up life could have arisen from the inorganic chemistry of the prebiotic earth was still just speculation. This changed in 1952 with an experiment conducted by Stanley Miller and his supervisor, Harold Urey.[16] In what is now known as the Miller-Urey experiment, these researchers evaporated water into an atmosphere of methane, ammonia, and hydrogen, gasses they believed might have been present in the atmosphere of the early earth. They provided a spark to mimic the electrical activity of lightning and cooled the atmosphere to create liquid. These researchers found that this simple process produced amino acids, the building blocks of life. This was a landmark finding that showed it was feasible for the foundations of organic chemistry to emerge out of the nonliving geochemistry of the earth before life became abundant. The actual composition of the gasses that were present at the origin of life were likely different from those used in this experiment, but despite this, the key takeaway still stands—it is possible for inorganic chemistry to become prebiotic organic chemistry through relatively simple reactions.

What do we know about the specific environments in which this transition from geochemistry to biochemistry might have taken place? Interestingly it is possible that these amino-acid precursors to life may have formed in outer space. For example, the amino acid tryptophan, which we consume in such foods as bananas and chocolate and which is only two chemical steps away from the serotonin in our brains, has been identified in deep interstellar space, specifically in a region of the Perseus molecular cloud where stars form.[17] The origins of life itself, however, are thought to have occurred on earth. Darwin speculated that life may have begun in a "warm little pond," while the biochemist Alexander Oparin suggested life had its origins in a "primordial soup" of intermixed chemicals.[18] Wherever life began, it needed not just the right chemistry but also a gradient of energy that it could dissipate. The kinds of life that we are most familiar with, such as plants and animals, ultimately all derive their energy from the sun. Plants photosynthesize, using the order of the sun to create their own order. Animals then consume the plants, using their borrowed solar order to organize their own bodies. Even carnivores consume secondhand solar order that has passed through plants and then other animals. The sun is not the only source of free energy that life can use to order itself, however, and it may have not been the source used by the first life-forms.

In 1977, scientists aboard a submersible called *Alvin* discovered ecosystems of living things around hydrothermal vents in the Galápagos Rift.[19] These organisms are termed extremophiles given the extreme conditions of the niches that they occupy. In 2000, the lost city hydrothermal field was identified on the Atlantic seafloor and has since become a favored location for the investigation of the origin of life.[20] These fissures in the ocean floor allow heated water to be expelled into the ocean, and they have a particular chemistry that may have been favorable to the emergence of the first life-forms.

One way to identify the specific environment life began in is to look for "metabolic fossils," metabolic pathways that are present in a wide range of living things, something we would expect of a particularly ancient pathway that would have been inherited by many life-forms. One candidate for a metabolic fossil is the reverse Krebs cycle.[21] In us, the Krebs cycle, also known as the citric-acid cycle, is responsible for breaking down carbon-based foods by oxygenating them in order to release their energy and, in doing so, releasing the by-products of carbon dioxide and water. In some bacteria, the cycle runs in the opposite direction, with carbon dioxide being used to create the carbon-based structures that support living systems. From our perspective, this version of the cycle appears to be running in reverse, but this is actually its original and more ancient version.

During respiration, the chemistry of the Krebs cycle is used to extract usable energy from food. In this process, electrons are stripped from food by proteins and are ultimately combined with oxygen and protons to form water. This current of electrons powers the movement of protons across the membrane, which in turn powers the creation of ATP, the cell's energetic currency. In this process disequilibrium, and therefore order, are being passed through different states. Let's consider how this plays out in creatures like us, who ultimately take our energy from the sun via plants. We start out with the energetic disequilibrium between the sun and space. This disequilibrium becomes the disequilibrium between the chemical potential energy that exists between the plant and its surroundings, into which this energy could be released. When consumed, the plant becomes food for us, and this energetic disequilibrium in the food powers proton disequilibrium in our cells, which in turn powers ATP synthesis, creating energy disequilibrium in the form of usable chemical energy inside our cells that can be used to do the work of maintaining living order. Here, the order of the sun has become the order of the living system that differentiates itself from its environment. Fundamentally, what is happening in this process is that forms are changing form rather than fully dissipating into a state of homogeneity.

In the reverse Krebs cycle, water and carbon dioxide are combined in order to manufacture the carbon compounds that make up the cell. One proposal for the specifics of the origins of terrestrial metabolism, which comes from the lab of University College London biochemist Nick Lane, is that pores in the minerals that make up hydrothermal vents provided a protective environment for a simple protometabolic chemistry similar to the reverse Krebs cycle to develop, before the advent of cell walls.[22] The interior of these pores are alkaline, while the exterior contains acidic seawater. This relative interior alkalinity and exterior acidity mirrors the structure of cells. This difference creates a proton gradient that facilitates the reaction of hydrogen and carbon dioxide, which can be used to make organic molecules. Ultimately, these products could be harnessed by other metabolic pathways to make sugars, fats, amino acids, and macromolecules. Once these self-sustaining metabolic pathways came into existence, selection could have honed them to add complexity, resulting in life as we know it.

SELECTION AND ADAPTIVE ORDER

Nothing makes sense in biology except in the light of evolution.
—THEODOSIUS DOBZHANSKY[23]

We have seen that life borrows order from other energy sources in order to exist, but this alone is not enough for a living thing to stay alive. Much like the sandcastle being hit by the waves, it is a statistical certainty that the motion of the world would fatally disorder us if we didn't take active steps to maintain our orderliness. If the sandcastle were to come alive and persist over time, it would be necessary for it to acquire the ability to navigate away from the waves in the service of its own survival and toward sources of energy so that it could perform the work of maintaining itself. All living things are processes of survival of this kind. Rather than being understood as a static orderly object, a living creature can be seen as a dance of self-persistence, a process rather than a thing.

Life arose when matter began dancing the dance of self-persistence. We can use the powerful logic of Darwinism to explain how this came about, in terms of general principles. Today, there is often a narrow focus on Darwinian selection only happening on genes that can mutate. Nothing was known of the existence of genes when Darwin was alive, however. What he was proposing was a far deeper principle. Anything that is good at persisting will persist—this is the essential logic that Darwin proposed. In *The Selfish Gene*, Richard Dawkins proposes that the idea of survival of the fittest is actually a specific case of a more fundamental principle, that of "survival of the stable."[24] The argument that things that are good at surviving will survive may seem tautological or even facile, but it is profound. This is an entirely naturalistic explanation for the forms of life we observe, an explanation whose logic is entirely self-contained and requires no outside intelligence to account for the incredibly complex living things that we see around us.

Darwin proposed this statistical logic to account for how it is that different species came into existence. We can think of evolution as instantiating a selection algorithm, a set of procedures to follow, that consists of three steps: vary, select, retain. In the case of the evolution of different species, the variation step consists in mutations occurring in the genes that give rise to different phenotypes. The selection step consists of organisms that are less good at surviving being more likely to die and therefore less likely to reproduce, while the ones that are well suited to their environment are more likely to survive and continue their line. The final step of retention occurs when the genes are passed on to the offspring of the creatures that survived this brutal selection process. This naturally occurring procedure is all that is needed to sculpt life into a diverse array of complex forms, from whales and olive trees to doves and corals.

This reasoning need not only apply to the process of speciation. Darwin's logic is equally applicable when it comes to explaining how life came into existence. Any system that succeeds in persisting will persist. The most general explanation of why life exists is simply that it was possible for matter to arrange itself in a self-sustaining process. Given enough time the logic of selection would take care of the rest. No matter how unlikely the conditions were for self-organization of this kind to arise, it only had to happen once for the process of persistence we call life to continue to this day. If we first imagine geochemical activity producing molecular networks of reactions that arise randomly, then we have the variation step. Once a simple self-perpetuating arrangement arose by chance, the process of selection by survival could occur, with all other chemical networks failing to perpetuate themselves. The retention of this capacity to self-perpetuate would be in the self-perpetuation itself until replication could arise. With further continual variation and pruning, more complex and adaptive forms could then gradually emerge.

The idea that selection acts in multiple places beyond the process of speciation has been referred to as universal Darwinism.[25] Darwin's proposal was that this process acts to shape the physical forms of different organisms. Giraffes came to have long necks, as this trait led to greater survival prospects in an environment where food was the leaves at the tops of tall trees. Through the selection algorithm, the problem of obtaining food in a given environment comes to be solved. Is this really the whole picture? Having a long neck isn't enough for a giraffe to consistently navigate toward food in a changing environment; it must also be able to adapt and learn within its lifetime. If the selection algorithm is so good at solving problems on this long timescale of multiple generations, then it would make sense for organisms to deploy this procedure to solve problems in their own lifetime, such as where to find food in changing conditions. We should expect evolution to equip living things with the ability to harness the power of selection for themselves.

This indeed appears to be what happened. In pioneering work on learning, B. F. Skinner suggested that the ability to learn through reinforcement, termed operant conditioning, shows a Darwinian logic of the kind described here. In his classic experiments, rats were rewarded with food when they pressed a lever. As a result, they came to press the rewarded lever more often.[26] While this learning may seem simple, this phenomenon demonstrates the way that organisms can adapt within their lifetime rather than depending on genetically hard-wired behavioral routines. Skinner argued that both evolution and learning of this kind were the result of a single principle of "selection by consequences."[27] In evolution, we see variation in genes and the resulting phenotypes, which is followed by selection through the struggle for survival and retention through successful reproduction, which allows the genes to continue to the next generation. During learning of this kind, there is variation in the initial behavior, with the rat pressing both levers. The selection occurs with certain behaviors being rewarded and retained as a result. The phrase *trial-and-error learning* captures this selectionist dynamic. We execute a variety of attempts (trial) to solve a problem, with the attempts that failed (error) being rejected and the behaviors that led to success being retained.

Gerald Edelman was a Nobel Prize–winning biologist and neuroscientist who argued that selectionist principles might explain what occurs in the brain to make this kind of learning possible. We are born with a vast number of connections between our neurons, and throughout development the ones that are best suited to our effective functioning in the world are retained. For Edelman, this is the initial selection stage that sculpts the structure of the brain. Next, a second stage of selection occurs as we navigate the world. According to this perspective, termed neural Darwinism, variation arises in the presence of a number of neural populations that could be recruited to solve a particular task.[28] When a population is recruited that leads to us successfully solving the problem at hand, the brain produces reward signals in the form of neuromodulators, such as dopamine, which enhance the strengths of the connections in this pathway. In this way, the appropriate population is selected and retained, leading to adaptive learning in one's lifetime.

We have a picture here where selection is operating across multiple scales of biological reality. It pulls organisms into existence, sculpts them to fit their niche, and selects genes that build the means of implementing selection-based learning algorithms within the lifetime of the organism. As living things, we are ourselves evolutionary processes that were created by a larger process of evolution. Evolution produced systems that themselves harness the power of selection because it is the most fundamental principle that allows complex forms to exist. Rather than conceiving of the emergence of life, evolution, and adaptation through learning as three separate phenomena, we can see them all as facets of the single phenomenon of selection that relentlessly operates to produce the complex forms of life across multiple scales.

I hope to have convinced you here of the incredible power of selection. Successful survival dynamics are the ones that survive over time, by definition, and any attempts that fail are brutally eradicated, and we are left only with forms that are good at surviving. Being good at surviving doesn't only happen on the intergenerational timescale of evolution; it also must happen moment by moment as you navigate your life. Systems that don't tend toward maintaining their order, that don't show such capacities as homeostasis and healing after damage, will not survive long enough to reproduce. Selection bequeaths us not only with bodily forms that are well suited to the niche that we inhabit but also with capacities of adaptation, repair, and the ability to navigate toward conditions that are conducive to life.

Given the lean logic of survival and selection, it should be no surprise that this relentless process has resulted in the robust survival phenomenon we call life coming into existence and proliferating all over the world. From the moment that the first chemical processes began engaging in self-perpetuation, the process of life never stopped. All the activity of life on earth, from plants growing to birds migrating to humans breathing, moving, and eating, is a continuation of this relentless process. In a sense, there is only one life-form, of which we are all part, one process of persistence that branches off into different organisms and species. You and any other living thing can be thought of as like the two hands of a person's body: relatively distinct but ultimately part of the same greater process.

EMERGENCE

If we view life through a reductionist lens and focus on chemical interactions alone, then we miss the organizational principles that define life. A focus on parts to the exclusion of wholes cannot explain how it is possible for collections of matter to crawl out of the ocean, climb up trees, fly across the sky, and even visit the moon. Something else is happening here that is not captured by our understanding of the physics of particles or of chemical reactions alone. Our mechanistic understanding of how it is that life came to exist represents an expansion of physics from the study of parts to the study of wholes, of organized systems that show novel behaviors that are more than the sum of their parts.

Take any living thing and rearrange its parts at random, and it most likely will no longer show the interesting behaviors associated with life. Random mixtures of carbon and other elements do not reproduce or seek out energy sources, yet when organized into life-forms, they seem to exhibit genuinely novel properties. To the reductionist, this apparent novelty is an illusion. Nothing new genuinely emerges with life. What appear to be novel properties are actually mere mirages, with physics doing all the heavy lifting. This issue is directly relevant to understanding the relationship between life and consciousness. Does consciousness really contribute to how we function as living beings, or is it an irrelevant epiphenomenon?

Considering such relationships takes us into the idea of causation. Things can only be said to exist in our scientific picture of the world if they take part in the unfolding of our universe; something that doesn't produce any effect on anything else cannot really be said to exist. If I claimed I possessed an object that was entirely hidden from reality, having no impact on it at all, then you would be right to question whether this supposed object could really be said to exist. For consciousness to be considered something that is genuinely part of our picture of the world, it must have some impact on the unfolding processes of nature.

A cause is an explanation of why something has happened. We can say something caused something else if the "something else" wouldn't have happened without the involvement of this cause.[29] We can say a billiard ball causes another ball to move, as without the first ball hitting the second, the second wouldn't have moved. In the reductionist worldview, the only dynamic that shapes the evolution of the universe is the impact of genuinely separate objects on each other. Aristotle had set out a system of different kinds of causality millennia earlier, and he termed this kind of cause efficient causation.[30] He believed there were also other causes at play in our universe, namely material, formal, and final causes. Material causation emphasizes the matter from which something is made. Formal causation captures the idea that the form of a thing can play a fundamental role in shaping its behavior. Final causation focuses on the purpose or end goal of a thing. It implies that objects and systems have an inherent purpose, or teleology, that guides their behavior and evolution. As patterns of self-persistence that are oriented toward the goal of survival, living things can be understood as showing a teleological causality of this kind.

To the reductionist, all of causality exists as efficient causation at the level of physics. Reality is just particles bumping into each other and affecting each other through the transfer of energy. From this perspective, every physical event in the universe is fully accounted for by the influence of other physical events. What would happen if we added additional efficient causes at an emergent level: for example, at the biological level of organization. This would violate the "closure of the physical," the idea that the physical world is fully self-contained—including additional efficient causes would mess up our neat picture of physics.[31] Extra causes also raise the issue of overdetermination, a situation where multiple causes are present that are each sufficient to produce a given effect. A circular causality in which biological wholes, such as organisms, are capable of influencing the very parts that they are made from creates a mess in this view. This mess is avoided by insisting on the unreality of the wholes. For the reductionist, a living thing is nothing more than the sum of its parts.

Philosopher Alicia Juarrero has offered another way of thinking about emergent causality that can make sense of the reality of wholes. In her picture, there is no issue with such emergent phenomena having their own causal properties that can influence the parts from which they are made. Juarrero conceives of this kind of causality in terms of constraint.[32] Consider a system that could be in a number of states. If it is completely free to occupy any of those states, then they are all equally probable. If something constrains the system, then it influences it to make a certain subset of states more likely than the other states. As a result, this constraining influence plays a role in sculpting the evolution of the system but without needing to transfer energy through efficient causation.

I explained earlier that the dynamics underpinning life involve catalytic closure, the enclosed circular form of the autocatalytic set that enables a system to make itself. Juarrero suggests that there are certain "enabling" constraints in the interactions of the molecules that make this emergent organization possible, but once it has emerged, the whole has its own regime of "governing" constraints that manage the ongoing operation of the system. In this way, we can account for the circular causality between parts and wholes. The V-shaped formation of a flock of birds comes into existence through the enabling constraints that each bird must be in a specific location with respect to the birds around it. Once it comes into existence, the ongoing V-shaped organization of the flock constrains how the birds taking part will move. In this way the parts and the whole come to influence each other harmoniously, where the emergent organization does nothing to directly interfere with the behavior of the parts that make it up but instead sculpts and guides their trajectories. As a result, the emergent layer governs the ongoing behavior of the system but never violates the smaller-scale enabling constraints. In this picture, constraints interlock to create coherence across all scales of reality.

Where and how did the interlocking constraints that make up living things arise? The wind does not appear to possess the right characteristics for this process of life to begin within it, for example, so what is it about our complex chemistry that makes the life process possible? In his book *A World beyond Physics*, complexity theorist Stuart Kauffman argues that there are three different ways in which autocatalytic sets of chemical reactions must loop back on themselves so that they can become self-reinforcing.[33] The first is catalytic task closure, the kind of self-producing circularity we have already seen, in which autocatalytic sets can recreate themselves, but there are two other kinds of closure at play. Living systems constrain the ways in which energy is released in a manner that allows work to be performed. This work is then used to create constraints that will allow further work to be performed, creating a constraint-work cycle, or feedback loop. For such a system to self-perpetuate, as life does, it needs to close back on itself so that all the necessary constraints are continually recreated and so all the work tasks are successfully performed. Proposed by Maël Montévil and Matteo Mossio, these two closures are termed "constraint closure" and "work-task closure."[34] This is the kind of governing constraint regime that came into existence with life and kept going over billions of years in an unbroken chain of continuity, up until today.

The self-creating circular structure of living things is what is referred to by the term *autopoiesis*. In autopoietic theory, organisms are conceived of as being inherently cognitive: that is, they are capable of sensing, deciding, and acting in a meaningful way. All living things must be coupled to their environment through a loop of sensation and action, according to this perspective, and the behavior of such a system must work to promote its continued autopoiesis. For an autopoietic system, certain features of the environment are valued as good (e.g., food), while others are valued as bad (e.g., physically harmful stimuli). This is an entirely objective description of how these features of the environment relate to the task of survival, however, and so does not imply any inner mental world on the part of the organism. It does point to a deep continuity between the life process and mentality, as explored in Evan Thompson's classic text *Mind in Life* and in *The Systems View of Life* by Fritjof Capra and Pier Luigi Luisi.[35] In addition to describing the physical processes that make up living things, protomental dynamics are necessary in order to fully describe life.

According to the evidence provided by consciousness science, your individual consciousness ends at the moment of death. The cotermination of mind and body in such moments points to a deep connection between experience and life. One possibility is that the living body acts as a power supply for the conscious brain, merely breaking down food for energy, so that a powerful supercomputer in our skulls can excrete experience through its awe-inspiring complexity and sheer information-processing might. Or perhaps a more subtle and elegant dance between life and consciousness is playing out as the stream of consciousness helps us navigate our journey from cradle to grave.

TWELVE

LIVING MIRRORS

You who want
knowledge,
seek the Oneness
Within

There you
will find
the clear mirror
already waiting
—HADEWIJCH II[1]

Ever since my mystical experience as a young teenager, I had been driven to understand what consciousness is. I'd spent the better half of a decade at Oxford, training to conduct neuroscientific research on the brain. I learned about how modern neuroscience could not conclusively determine how experience relates to the brain and how we had hit a dead end in trying to understand where consciousness fits into our broader understanding of the world. I didn't know if we would ever find a solution to these problems. Then the insight that resolved my search came to me one day in a flash.

That morning, I'd cruised my houseboat along the canals to London's picturesque Little Venice, surrounded by the teeming ecology of the waterways on my journey to this idyllic setting. Now, I was moored up amid the weeds and willows. I lay down on the narrow sofa inside the boat, and my mind returned to that perennial mystery of consciousness. I began thinking about the history of our universe so that I could place the evolution of the brain in a wider context. Closing my eyes, I imagined myself to be the evolving universe, developing from its origin to us.

I started with the bits with which we are all familiar: the big bang, planets forming, and the occurrence of the first life-forms on earth. My imagination struggled at first, trying to grasp what it would feel like to be the expanding lifeless cosmos. Then I got to the origin of life. I felt the web of matter that makes up the universe fold in on itself to create a little, enclosed bubble: the first single-celled organism. I felt how, in order to keep itself together, this organism needed to consistently interface with the world around it in a way that reached beyond its boundaries to anticipate what was going on outside itself to successfully navigate in the world. I sat bolt upright. Was that it? Was that the moment consciousness came into existence, when orderly living things attempted to separate from the disorderly world around them? Could the focus on the human brain have been a distraction created by our own hubris?

Before that fateful day, I was fully convinced that the brain was the place to look for the physical basis of consciousness, even though I was unsure if we would ever succeed in this search. I'd trained as a neuroscientist for this reason. The mainstream of consciousness science generally takes it as a given that the brain, or at least the nervous system, is responsible for bringing experience into existence. The possibility that this idea might be mistaken is not something I was expecting. Given the evidence provided in earlier chapters on the neuroscience of human consciousness, the idea that the brain generates experience might seem to be on solid ground, despite philosophical issues, such as the hard problem. It is worth considering, however, whether the involvement of the brain in *our* consciousness necessarily implies anything about where and how this phenomenon came into existence. The brain may simply elaborate on a process that already existed before nervous systems evolved.

If consciousness came into existence with the first life-forms, might all living things be capable of experience? As I walked past the tall plane trees on Tavistock Square on my way to the lab, I wondered if it might feel like something to be these majestic elephantine organisms. Surely they would have no reason to feel my hand on their bark, but could their sensing of moisture levels in the soil and the subsequent moving of their roots toward the water feel like something? After days of pondering this issue, I decided that, as odd as it seemed, there was nothing that made it impossible for these trees to feel the world around them. Clearly natural forms were capable of being conscious; I was such a form, and I was conscious. Believing consciousness to be a product of the brain alone, however, raised insoluble riddles like the hard problem. Seeing experience as pervading the living world was more plausible and less problematic—and it subverted the hard problem altogether.

KNOWLEDGE IS ORDER

We have seen that living things emerge as islands of order in a universe moving toward intermixing and disarray. One way to internally maintain order is through self-maintaining feedback processes of self-creation. This is only part of the story, however. The environment is not a placid landscape free of challenges for life. On the contrary, moment by moment, the second law creeps up on the living creature, tugging at its boundaries, sucking it toward the chaos from which it emerged.

At the start of this book, I explore how it is that we mistakenly see separation as genuinely existing rather than appreciating it for the mental construct that it is. Living things attempt separation from the surrounding environment, but they never succeed. It is impossible to fully separate from the universe for even a moment. If we are to fully understand life, then we must appreciate that organism and environment are two sides of the same coin. The phenomenon of life is the phenomenon of *living in an environment*, not of genuinely separate creatures that are set apart from the world.

With this in mind, consider how it is that any organism prevents itself from succumbing to the destructive forces that exist outside itself. It must be able to take actions to avoid danger, and to do this it must be able to predict the danger, and for this it must have some way of sensing what is occurring in the world. The same is true for exploiting opportunities that promote survival. In creatures where this process was successful enough for them to survive, there must be some correspondence between the living thing's predictions about the dangers and opportunities of the world and the actual state of the world. This kind of correspondence is what we call information. Living things are necessarily informational systems.

What is information? This term can have multiple meanings, but it was first rigorously defined by the engineer Claude Shannon in 1948.[2] Something carries information about something else if examining the former reduces our uncertainty about the state of the latter. For example, if you ask me where an object is and I tell you the truth, then we can say that there is information in my communication, as it successfully reduces your uncertainty about the object's location. In the same way, the internal states of living things that successfully survive after a brutal selection process must do so by forming models that carry information about the environment, information that can be leveraged to achieve successful navigation in the world in the service of self-persistence.

Surviving over time means locally reducing one's entropy in order to maintain a region of order. This necessarily involves obtaining information about the environment, as without such information, a system could not behave in such a way as to stay orderly. Does that mean entropy and information are related in some way? When Shannon showed his formula for calculating information to famed physicist John von Neumann, von Neumann pointed out that it was the same formula as the one for quantifying entropy. For organisms, attempting to acquire knowledge of the world is inherently part of the project of maintaining their order. It really is true that knowledge is power—for the organism, knowledge of the world is the power to stay alive.

How are information and entropy related, precisely? We can think of these as two ways of looking at a given system: one in terms of physical organization (entropy) and the other in terms of uncertainty and knowledge (information). Consider a disorderly system that can exist in many different states: for example, one hundred oxygen molecules in a glass box that are randomly arranged. If we look at it through the lens of its physical organization, we can say it is high in entropy because there are many states the system can occupy. The oxygen molecules can be absolutely anywhere if they are truly randomly arranged.

Now look at the system through the lens of information theory, understanding that information is uncertainty reduction. In this system there is very high uncertainty about what precise state the one hundred molecules are in before they are observed because they could be in any number of possible arrangements. An observation of the actual state of the system is therefore highly informative, as it dramatically reduces the uncertainty of it being in any one of many different states to the single actual state that is observed. Observing a highly entropic system gives us a lot of information.

Now consider the same box where the oxygen molecules are all forced to be on a single surface of the container. The entropy of the system is now lower than before because there are fewer states that the molecules can be in, compared to when they were free floating. Now, when we make an observation of the system, we receive less information than before because we have less uncertainty about the arrangement of the particles before making the observation. If we have less uncertainty to reduce, then we can't receive as much information. A system that stays orderly, through the physical organization lens, is a system that exists in a restricted number of possible states—it is a system that has low entropy. Through the lens of information theory, such a system has low uncertainty; therefore, the events associated with its operation will be low in information. A system that keeps its entropy low over time is a system that minimizes the information it receives.

It may sound strange that living systems and brains act to minimize information, given that we think of them as information-processing systems. We celebrate the large information-processing capacities of our computers, and if we subconsciously are seeking a flattering technological analogy for the human brain that could justify human exceptionalism, then the idea that an effective brain minimizes the information it receives may feel counterintuitive. Receiving minimal new information is not the same thing as being uninformed, however—quite the opposite. It is precisely because the system functions in such a way as to correctly anticipate the state of the world that it ends up not being surprised. It is because a successful living system is so well informed about the world that it manages to receive minimal new information.

Is this connection really necessary? Couldn't you have a system that keeps itself orderly over time while not minimizing the information it receives? Systems that minimize their entropy, through the lens of physical organization, *are* systems that minimize the informativeness of unfolding events, through the lens of information theory. This necessary link exists because these are two descriptions of the same phenomenon. Minimizing information in this way is not just one strategy for survival; it is the same thing as survival.

What components are necessary for an information-processing system? A good example of such a system is a thermostat. The thermostat senses the air temperature, then switches on or off a heater, depending on the information it receives and whether it is higher or lower than a specific target temperature. Such a system has the ability to sense something outside itself, temperature in this case, and to respond with an action, turning the heater on or off. Self-organizing systems that succeed in existing over time, like life, also exhibit the capacity to sense the world and to act in response. In contrast to a thermostat, such a self-organizing system cannot simply wait for the situation to change and then respond. It must actively anticipate the events that could disrupt its organization.

I presented the free-energy principle (FEP) earlier as a precise formulation of how brains can approximate Bayesian inference, the process through which we leverage prior knowledge in combination with sensory evidence to perceive the world around us. I showed that it was impossible for the brain to perform true Bayesian inference, as it would need to know all the possible causes for a given piece of sensory evidence, something it has no way of knowing. The FEP provided a pragmatic workaround so this issue was avoided. The brain could instead start with any belief regarding what might be giving rise to the sensory input and simply update this belief in line with how surprising the next piece of sensory evidence is if that belief were correct. If you receive exactly the sensory inputs you expect to receive, then there is every reason to think your belief is accurate, and given that the input is not surprising, there is no need to update your belief. If the sensory evidence you receive is totally inconsistent with your current belief, then this evidence is very surprising, so you'd better dramatically alter your belief about what is going on in the wider world because it most likely isn't very accurate.

The FEP began as a theory of predictive processing in brains, but it was then expanded to describe exactly the kind of organized information-processing systems considered here. Karl Friston, the creator of the FEP, has shown that, in such a system, the internal states can be interpreted as a model of events occurring in the world.[3] Specifically, they can be seen as Bayesian beliefs of the hidden causes that give rise to sensory events—a picture of what is going on in the wider world. Now we can see how staying orderly is linked to information. A low-entropy system is a system that succeeds in not being surprised by unfolding events; it is a system that successfully predicts the world around it, that therefore minimizes how informative unfolding events are. No news is good news from the perspective of life.

In the language of information theory, this process is referred to as minimizing surprise. In Bayesian statistics, the same process is seen as a system maximizing the evidence for its model of the world. Staying orderly is only possible if one is generally not surprised by ongoing events, and this is only possible if one builds up an accurate model of the world. Formally, these seemingly separate processes are the same phenomenon. Living things must necessarily build up and maintain models of what is going on in the world for the entire duration of their existence. We can think of these models as creating a simulation of the world and oneself within it. I believe that this simulation is what we call consciousness.

Let us forget all we have seen about the specifics of the origins of life. The central insight here is that the only systems that could possibly have made it through all the relevant selection filters to get to survive as living things today are systems that attempt to know the world around them, that simulate the world and themselves in it to successfully navigate toward their goals. We cannot say the same thing of nonliving things, such as rocks or the wind. To be a living thing is necessarily to be a conscious thing.[4]

BEYOND NEUROCHAUVINISM

The FEP shows how predictive processing is a crucial factor in understanding how it is that matter became alive and continues to survive. I have presented the science that shows human experience to be based in exactly these predictive principles. The common assumption is that experience is somehow added on to the survival processes that the body is engaged in, opening up the possibility that unconscious living things could have come first, with consciousness emerging later. We see here that the predictive dynamics that underpin experience are not a recent evolutionary add on but are a fundamental part of life. We have no basis to restrict consciousness to brains alone, and we should not, as doing so raises insoluble philosophical issues in the form of the hard problem. Saying that a position raises "insoluble philosophical issues" is a polite way of saying that we know that way of thinking to be incorrect, so it must be abandoned.

In the brain-centric approach, consciousness is often conceived of as a disembodied substrate-independent computation. From the perspective of organizational functionalism, emergent phenomena like consciousness can only exist through systematic modes of organization, and in our case, that means through our holistic organization as living things. Taking these points together, we can see a clear moment in the evolution of the universe at which consciousness emerged—the moment life came into existence.

Consciousness could have emerged with the novel organizational mode that we call life, but no comparable level of emergence has occurred throughout the history of biological evolution. We do not have a biology of mammals that is separate from the biology of bacteria. It's all biology because it all exists as a single kind of emergent phenomenon. We have no basis for saying that consciousness emerged through evolution and every reason to believe that we are conscious because we are alive. Being alive gives us the capacity to experience—brains are a later add-on.

BEING IS BELIEVING

The origin of life is also the origin of systems that attempt to know the world. To be alive is to be engaged in epistemic—that is, knowledge-gathering—behavior. We must be careful with the word *knowledge* here, however, as living things do not seek objective truths but rather information regarding patterns that are relevant for their own survival. I experience maggot-ridden food as disgusting, while the maggots may experience it as delicious. Neither of these attitudes reflect knowledge of the true nature of the food because there is no true objective state regarding its being appetizing or not. Instead, there are only the relational dynamics between the food and the organism in question. If I eat it, I will become sick, so for me the "truth" of the situation is that it is "bad" food, while for the maggots it is a valid source of nutrition. There is no true knowledge of the goodness or badness of the food to be gathered here, independent from the perspectives from which these qualities are being assessed.

According to Plato, knowledge is justified true belief, making it a subcategory of belief.[5] Rather than thinking of organisms gaining knowledge of the true state of the world, we should be thinking in terms of them believing things about the world. The term *belief* offers a useful way of closing the gap between first- and third-person descriptions of consciousness. Earlier I mentioned that Kant argued that we never know the world directly but only experience our phenomenal representations of it. The appearances that arise in awareness do not reflect the way the world is, they are simply ways we believe the world to be. From this Kantian perspective, we can think of the problem of consciousness as one of describing, in objective terms, how it is that certain systems come to have the ability to believe in a world consisting of qualities. It is because we have this capacity for belief that we are conscious.

The use of the term *belief* here should not be seen as being limited to conceptual beliefs that can be stated in words, such as the belief that penguins live at the South Pole. We can think of believing as being, at the most fundamental level, the ability to hold the world to be a certain way. When you consciously perceive an object in front of you, you wordlessly believe there to be an object existing in a world, despite it being the case that no separate object with the qualitative traits you perceive actually exists in the objective description of our universe.

In addition to this subjective description of the contents of consciousness as beliefs, we can consider the predictive models that living systems instantiate as beliefs in the state of the world. In Bayesian theory, a hypothesis about the state of the world is already known as a belief. This term has come to be identified with mere mathematical distributions in probability theory. In embodied living systems, however, the mathematics of belief updating has been found to be a great model for the experiential beliefs that make up consciousness. For this reason, I feel it to be a useful term in accounting for the physical dynamics that underpin consciousness.

The use of this term could invite confusion between the mathematical model and the target phenomenon of experience, although this is easily avoided by understanding the difference between the map (the mathematics) and the territory (consciousness). Still, I use this term because I feel it to be no coincidence that the word *belief* can be used in both cases. We are systems capable of belief—that is what makes us conscious. It is also the reason that a living human came up with Bayesian theory; a rock would not be capable of theorizing about belief updating in light of evidence because this is not something it is capable of. I use the word *belief* because it points to how we can close the gap between our quantitative understanding of the operation of the world, specifically the mathematics and physics of believing systems, and our experience *as* such systems.

BEING A LIVING MIRROR

The specific dynamics described by the FEP involve a consideration of the functional boundary that living things must assert against the rest of the world to maintain their order. Technically, this boundary is referred to as a Markov blanket (named after the mathematician Andrey Markov), but it is simply the parts of the system that isolate it from its environment. This boundary has two components: parts that can sense the outside world and parts that can act in response. Structures inside the system are then responsible for mediating from sensation to action, and it is these internal structures that turn out to necessarily carry information about the hidden causes behind the sensory input in systems that succeed in surviving over time. As a result of its survival dynamics, the organism builds up a simulation of what is going on in the wider world, an inherently imaginative act that goes beyond the sensory data it receives. The organism attempts to know what is going on in its environment so it can survive within it.

Organisms necessarily believe things about the world beyond their boundaries so they can survive. Recall that these beliefs are guesses or hypotheses about the state of the world rather than totally accurate knowledge. Instead of being encoded representations of things in the environment, these beliefs are generated from within. If such beliefs were entirely unrelated to the patterns of the outside world, however, the creature holding these beliefs wouldn't last for long. Survival sculpts these beliefs through the process of surprise minimization captured by the FEP so that they come to mirror useful patterns in the environment. I originally published this theory of consciousness under the name the living mirror theory to capture this reflective dynamic and its necessary connection to life.

There is more to this dynamic than the simple reflection of external patterns, however. The patterns of the world that come to be reflected in the beliefs that a living system holds are the ones that are relevant to the survival of the organism; the pattern always includes the organism. In the case of perceiving rotten food, you are not only perceiving its physical state of decay, but you also experience it as viscerally disgusting because the relevant pattern for your survival includes the negative effects of what would happen if you were to eat it. A different pattern exists for a maggot, as eating the same food would presumably not have the same negative impact on its survival. A complex survival-relevant pattern exists in the scene that involves the state of the food, its location, and the effect it would have on you if you ate it. It is this pattern that becomes reflected in the belief that disgusting food exists in front of you in that moment. Holding this belief is synonymous with the pheromonal experience of that food existing and being disgusting. Outside our capacity to hold such relational beliefs, disgusting food does not objectively exist in our third-person description of the world.

To complicate things further, it is not only objects in the outside world that the organism can hold beliefs about. Living things continually act on the world, allowing them to infer their own existence from the sensory signals that arise from their actions. When we think of living things as constructing beliefs about the world, we must appreciate that this refers to not just the outside environment from the organism's perspective but also the whole scene, including the organism. The boundary of a living system is an informational mirror that reflects any patterns that are relevant to its survival that can be inferred from the sensory signals it receives, including the organism's own presence in the scene.

To see how this whole process plays out, imagine a population of early single-celled organisms shortly after life got started on earth. Just north of the hydrothermal vent that they call home is a fissure in the ocean floor through which red-hot magma pours out into the water. Random variations occur within this population of organisms, and some are not well equipped to detect this dangerous area of molten rock, so they drift into it and do not survive. Their entropy is mortally increased, their boundary dissolves, and they meld once again with wider reality. Through chance variations, some develop heat sensors on their membrane that allow them to detect when they are getting too near the lava. Others also develop photoreceptors that are sensitive to light. Through the brutal process of selection, the organisms that end up surviving for any length of time are the ones that have developed the ability to leverage all the information available to build up a simulation of what is going on in the world around them. Remember that this is not optional for survival. To keep one's entropy low, one must necessarily infer the state of the world around oneself.

The molten rock preferentially reflects long wavelengths of light, and this correlation between light wavelengths and heat proves a useful piece of information for our little creatures in identifying areas of the sea floor that may be dangerously hot. One of the successful organisms not only ends up with simple representations of the temperature of wavelengths of light but also carries a hypothesis or belief about what is going on in the world that could be giving rise to these sensory signals. A complex statistical pattern exists in this scene that involves not only correlations between different physical features of the lava and the surroundings, such as the thermal and visual reflectance properties, but also the way in which they relate to the survival of the organism. It is this pattern that is captured in the belief that the organism comes to form, a belief that might feel like the experience of there being dangerous red-hot lava in front of the organism, if this simple creature were able to put words to this experience. Through the process of self-persistence, these survival-relevant patterns come to be reflected in the beliefs that the system holds.

This explanation of consciousness may be more compelling if we recall the metaphysical picture of reality that I unpack in part I. In that context, this perspective could be called biopsychic Darwinian transcendental dialectical process-relational neutral monism, but I think *living mirror theory* is more catchy. Let's take a moment to unpack the different facets of this perspective that are captured by this cumbersome phrase. *Biopsychic* refers to the idea that all living things are conscious. *Darwinian* refers to the core role of selection dynamics in producing the Bayesian beliefs that underlie the contents of consciousness. *Transcendental* points to the acknowledgment of a world in itself that exists beyond our individual subjective experiences, as well as the way in which our subjective experiences are shaped by the structure of our own minds. *Dialectical* refers to the metaphysical interplay of formlessness and form, being and nonbeing, explored in chapter 4. *Process-relational* refers to the idea that the forms of reality exist as interdependent dynamics of change rather than consisting of truly separate objects that rest on a ground of solid substances. *Neutral monism* is the idea that at the foundation of reality is neither matter nor mind (I claim it is the dialectical interplay of formlessness and process-relational forms), with matter and minds being different dynamics that arise within this neutral ground. Taken together, these descriptions point to a unified picture of reality where there is no fundamental divide between the relations we call matter and those we call consciousness.

THE LIGHT OF AWARENESS

In the introduction, I explore the idea that awareness lies at the core of the mystery of consciousness. Why doesn't our simulation of the world simply occur in the dark, through nonconscious physical information processing? Why is it illuminated? Why is there this quality of "knowingness" in experience?[6] When I look at the blue of the sky, there is a knowing of that blue quality. If a camera captures the same visual input, there is presumably no knowingness. As Kant suggested, we cannot directly know the world in itself, but we can believe things about it. The phenomenal world of experience is made not of knowing but of belief. The "knowingness" quality that we can identify in experience can be more accurately understood as a "believingness." The space of awareness is a belief space. To experience oneself as existing in a world is to believe that one is existing in a world. In the picture presented here, the light of awareness is the capacity of organisms to hold beliefs.

The survival-relevant patterns that are reflected in the functioning of the organism arise in this belief space. We can examine this belief space both subjectively and objectively. Scientists routinely deal in abstract spaces in the course of their research. For example, if you repeatedly roll two dice, you could plot the results on a 2D graph, and we could think about this space of all the possible combinations of numbers that could arise from rolling two dice: 1 and 1, 1 and 2, 6 and 6, and so on. Here we have a "two-dice throw result space." Similar spaces can be constructed for any other physical process we might study, from neural activity to gene expression. The beliefs that living systems hold about the hidden causes behind sensory events can be conceived of in the same way—they form a belief space.

From the perspective of the organism, this belief space can be seen as the nondual space of formless awareness. This is not the same kind of space as physical space. It is as if a new layer of existence opens up with the emergent dynamics that make life and consciousness possible. Imagine a bubble or a ball-shaped mirror. The surface of such a ball or bubble casts a reflection of the entire scene around it. Put two such objects side by side, and each appears to contain an entire world. Organisms function as a kind of surface that casts reflections of survival-relevant patterns in the world that are made of beliefs. These beliefs are simply relational dynamics, so in the same way that the pattern in the mirror and what it reflects in the outside world are both patterns, both the physical world and consciousness are relational dynamics.

This belief space can also be conceived of as what philosopher Tomas Metzinger has called an epistemic space.[7] Metzinger holds that formless awareness, or pure consciousness, as he calls it, arises as the result of the brain representing this epistemic space. I suggest that we do away with the representation stage and instead see awareness as the epistemic space itself, the space in which beliefs about the qualitative character of the world can arise.

Given the inferential dynamics of living systems, we can have no basis to argue that this epistemic space emerges at any point during evolution. Instead, we must see that it came into existence through the interactions between organism and environment that are essential for survival. No matter the qualitative nature of the experience of nonhuman creatures, we have every reason to believe the capacity to experience to have come into existence with the origins of life.

WHERE DO QUALITIES EXIST?

Experience is never limited, and it is never complete; it is an immense sensibility, a kind of huge spider-web of the finest silken threads suspended in the chamber of consciousness, and catching every air-borne particle in its tissue.—HENRY JAMES[8]

What exactly is a qualitative experience? Take a simple experience, such as seeing the blue of the sky. How does this map onto the physical operations of a living organism to bridge the mind-matter gap? We need to understand that an experience is not something that can be directly connected to a single mechanism in the organism, as is the aim of the reductionist project—the experience of blue alone cannot be accounted for by the presence of a blue-experience-generating mechanism in the organism. We must instead understand what in the physical dynamics of the organism *is* the conscious experience. This takes us back to the idea of an identity relationship, explored in chapter 9, but we must look beyond the idea of consciousness being identical to a specific mechanistic process in the brain.

We can think of consciousness as a *process* of believing rather than as an objectifiable thing or substance. As such, we cannot locate consciousness in the physical structure of a living system but must consider it as a mode of operation, a way of behaving. This accounts for the seemingly immaterial nature of consciousness, as it is an epistemic behavior that is associated with the whole physical organism as it relates to its environment. If we were to ask where consciousness is located, it would be like asking, "Where is running located in a jogger?" We might be tempted to say that running is located in the legs, but without taking into account the surface beneath the jogger, gravity, their circulatory system, their senses, and more, we wouldn't have a full description of how it is that this behavior exists. Consciousness exists in the systemic organization of an organism, not as something additional or objectifiable that is somehow different from the operation of the physical body.

I have shown that this perspective can synthesize aspects of idealism, panpsychism, and physicalism into a single worldview, and through the point being made here, we can also incorporate the seemingly incompatible stance of illusionism into this synthesis. The "illusionist" is a monist who argues that consciousness is nothing separate from the operation of the physical world, as I am arguing here. Rather than calling consciousness an illusion, however, I would say that it is a process that is empty of any independent essence, like everything else in existence. Like all phenomena, consciousness only exists as a certain kind of interaction within the processes that make up reality. Much as money doesn't have any inherent value apart from the way it is used in our economic relations, consciousness has no existence apart from the relations between organism and environment. Experience is not something different from the functioning of the organism. The act of alternately moving your legs to transport yourself across a room does not generate something different called walking; it *is* walking. There is no hard problem of walking created by dissociating these two levels of description. If we consider consciousness to exist as the organized interaction between the whole organism and its environment rather than as something separate from a mechanical unconscious body, then there is no longer a hard problem of consciousness.

Now we can address the question of how it is possible for a qualitative experience to exist in a world of quantitative matter. Rather than seeking to map a specific experience to a specific quantitative mechanism, we must see that the qualitative character of a single experience is defined against other experiences, not against the physical world. Consider how an emergent relational structure like an ecology exists. We do not explain components of the ecology with appeals to the individual animals that make it up. We look at the relations between the creatures in the ecosystem—an interdependent structure. In the same way, we can conceive of the contents of consciousness as consisting of an interdependent framework of beliefs in qualities.

I explained earlier that structures of this kind can emerge through a process of relational codefining, and we can conceive of this process as underlying the existence of qualitative experience. Red is red because it is not blue or any other color, for example. A world in which we could see only red would be the same as a world in which there is no color at all. An individual qualitative experience such as the redness of red is empty of any independent existence. Only the framework of codefined qualitative beliefs exists, and it exists through the emergent systemic organization of living systems as they engage with their environment.

What does it mean for something to be qualitative? Quantification occurs when we can measure the amount or extent of a phenomenon. The issue of how good versus bad for one's survival something is, however, is not quantifiable. Rather than being able to measure how many units of survival we are currently displaying, we can only consider a continuum from success to failure. The qualitative pleasure of drinking water when thirsty captures a move in the direction of survival and thriving, while the qualitative pain as a result of severe thirst captures movement in the opposite direction. Pain has the quality of being negative in light of its impact on your prospects of survival.

We could say that quantities are absolute, whereas qualities are relative. If I said a given object weighs two hundred pounds, that would be a statement of a quantitative property of that object. If I asked you if it possessed the quality of heaviness, this would be a relative judgment. An object cannot be heavy in any absolute sense. If it's a bag of sugar, then you might say it has the quality of heaviness compared to other bags of sugar that you typically come across in the supermarket. If it was a new kind of car, then you would say it was exceptionally light. Qualities are also relative to a wider context. If you're asking me to lift the two-hundred-pound car, then it does have the quality of heaviness in that context.

We can ask, however, whether quantities are truly absolute and free of both relationality and contextuality. The quantity in this example, the weight of two hundred pounds, is only meaningful given the context of it being measured by an earthbound creature where gravity is acting on both the object and the individual weighing it. I explain in chapter 5 that matter itself appears to be relational. That is, even a physical particle is a relative event. It turns out that we are best placed to close the gap between absolute quantity and relative quality not by explaining how it is that qualities can exist in a quantitative world but by appreciating that absolute quantities are actually relative qualities in disguise. They only look absolute when we do not factor in the context, including ourselves, within which they are being examined. This brings us back to the relational metaphysics in chapters 4 and 5.

I've shown how science tells us that we do not experience reality directly but are instead imprisoned in a controlled hallucination of the world around us and of ourselves within it. While this is true, the exploration of how qualities come into existence provided here can recover our intuition that we really are in touch with the world in some sense. While pain does not exist in the objective description of the universe, it does exist as a relational phenomenon from your perspective. As a living thing, for you it is genuinely true that bodily harm is bad, for example. While the contents of our experience are not real on an absolute, objective level, they are real on the relative level. This dual reality is quite empowering.

If you are troubled by a weed in your garden, then you can be reassured that it is reasonable for you to perceive it as a weed; you are not fundamentally wrong and disconnected from reality just because "weedness" doesn't exist in the physical description of the world. It is a weed to you simply because you don't want it in your garden. It is also true that this plant is not objectively a weed. It only appears so given how you are relating to it. You can choose to relate to it in another way. As a result, you can undercut the belief that it is a weed if it is causing you distress and appreciate that your way of looking at the plant is not fixed. You can choose to appreciate it as a plant in its own right without rejecting it, and you may find that its presence causes you less suffering as a result.

LIVING AND FEELING

I've covered a lot of terrain so far, and it has all been in the service of making what ultimately amounts to a ridiculously simple claim about why consciousness exists. When you were a child, before you knew anything about brains or consciousness, what answer would you have given if someone had asked you why a bird sees? Without knowledge of brains, you might have answered that it sees so it knows where it's going as it flies, so that it can find worms to eat and avoid dangerous animals. To summarize this claim, the bird sees so that it can engage in survival-related behavior. Even if you said that it sees because it has eyes, a follow-up question about why it has eyes would most likely have elicited an answer of this kind. My contention is that this intuition is fundamentally correct and applies to all living things. We experience because we need to in order to stay alive. Only systems capable of experience could have successfully survived to exist today. It is only our cultural commitments to such assumptions as human exceptionalism; neurocentrism; and our unexamined folk-psychological notions, like belief in substance, self, and separation, that make this issue seem more complicated than it needs to be.

The claim here is that consciousness is an inherent part of being alive. Let's recap why this should be the case. To survive in a universe that is moving toward disorder, it is not enough for an organism to simply sense and respond to the world around it; it must anticipate and simulate the world outside itself. Imagine crossing the road with your eyes closed and deciding on your strategy of moving out of the way of vehicles by sensing their touch on your skin and only then responding. You wouldn't last long. Instead, the strategy that life had to hit upon in order to survive over time was to infer the patterns of the world that explained the sensory information it received and then to take actions in line with maintaining its order over time. In our universe, maintaining order necessarily involves inferring the state of the world, and it is this process of simulating a world that we call consciousness.

At this point, objections may be arising in your mind. Rest assured; I take time later to explore how feasible this seemingly radical conclusion really is. I also return later to our philosophical explorations of the first few chapters to see why biopsychism makes sense in light of the metaphysical picture explored there. For now, all I ask is that you maintain an open mind.

THIRTEEN

IS CONSCIOUSNESS POSSIBLE WITHOUT BRAINS?

From Bacteria to Bots

It is hardly an exaggeration to say that the tip of the radicle (root) thus endowed, and having the power of directing the movements of the adjoining parts, acts like the brain of one of the lower animals; the brain being seated within the anterior end of the body, receiving impressions from the sense-organs, and directing the several movements.
—CHARLES DARWIN[1]

THE BIOPSYCHISM TABOO

For such a simple idea, the position of biopsychism (i.e., that all living things are conscious) has been exceptionally rare in Western thought. The situation is different in many Eastern philosophical traditions, where Buddhism, for example, makes it clear that all living things can feel and suffer. The enactive perspective explored in chapter 10, in connection with the concept of autopoiesis and the link between cognition and the embodied organism, was informed by Buddhist philosophy, yet key thinkers did not accept this stance of all living things being conscious—biopsychism was rejected here, too. While successfully making a link between life and mind, this autopoietic approach initially confined itself to the objective study of organisms and observable cognitive processes instead of attempting to close the gap between matter and experience. Where consciousness was considered in this framework, it was typically in the context of the embodied brain.

Evan Thompson, a key figure in the enactive approach, initially suspected that the life process was not enough for consciousness and that experience came into existence with nervous systems.[2] In 2022, he revised his views on this topic, however, to reconsider biopsychism.[3] Thompson came to the conclusion that the idea of all life being conscious was a philosophically plausible position after all, although he felt that the gap between matter and mind had not been successfully closed by existing work in this area. Thompson argued that the enactive conception of life does involve the emergence of value with respect to the organism in this view (e.g., nutritious food is "good"; bodily harm is "bad") but also that the *experience* of value is something else, something not accounted for in the third-person description of living things provided by the theory of autopoiesis.

Thompson had been influenced by Lynn Margulis, the visionary evolutionary biologist who first proposed that our cells are symbiotes, with the mitochondria that power them once existing as independent bacterial organisms before they were absorbed by larger single-celled creatures. In 1995, Margulis coauthored a book with her son Dorion Sagan titled *What Is Life?* In it the authors state their belief that "not just animals are conscious, but every organic being, every autopoietic cell is conscious."[4] Margulis's theory of symbiosis is a relational one based in understanding the interdependent collaboration between organisms that gives rise to a new whole, in the form of our cells. I think it is no coincidence that a relational thinker of this kind also intuited the possibility of consciousness existing in all living things.

Another thinker who has tried to move beyond the brain-centric approach by arguing that consciousness came into existence with the first simple organisms is the psychologist Arthur Reber. Reber is best known for his work on implicit learning, where information is acquired unintentionally without the individual being consciously aware of it. One day he was observing a caterpillar eating a basil plant, and he was struck by the feeling that it, too, may experience the world. Caterpillars do have nervous systems, so this may not seem like a radical departure from the mainstream, but it led Reber down a path to concluding that consciousness depends not on nervous systems but on flexible cell walls, the ability to sense the environment, and the ability to locomote. He called this the cellular basis of consciousness theory.[5] While bacteria would be conscious in this framework, it initially ruled out plants as being capable of supporting experience. As a result, it was not a truly biopsychist position, as certain forms of life were excluded from being conscious. He would come to reconsider this view, however, after collaborating with researchers in the field of plant intelligence, such as František Baluška.[6]

In 2016, the journal *Animal Sentience* allowed submissions for peer commentaries from other academics on Reber's proposal. It received pushback, with commentators arguing that it failed to close the gap between matter and mind. In response to these commentaries, Reber agreed that the emergence of consciousness from life was indeed still mysterious and something of a "miracle," pointing to the same gap that Thompson had confronted.[7] My contention is that this gap was felt to exist as a result of the common metaphysical assumptions that we have operating in the background. With a process-relational metaphysical framework (explored in chapter 4) and the perspective of nonreductive organizational functionalism outlined in this book (in chapter 10), I believe no such miracle is required, and no gap remains between the physical world and consciousness.

Some thinkers have seen the life process not as the wellspring of consciousness itself but as the necessary foundation on which conscious brains rest. One example can be found in the work of the neuroanthropologist Terence Deacon. In his work *Incomplete Nature*, Deacon lays out a wonderfully detailed account of how thermodynamic constraints lead to the emergence of the self-perpetuating behavior that characterizes life via an intermediate step that shows how simple structures emerge in some thermodynamic systems. Deacon argues that the self-perpetuating nature of life is responsible for the ability to sense and respond to the world, with consciousness emerging in animals that use their nervous systems to store information in order to guide action, moving them beyond mere unconscious reflex.[8] In this view, consciousness is ultimately grounded in life but requires nervous systems to come into existence.

A similar stance is held by Mark Bickhard, whose interactivist model sees single-celled organisms as responding by simple reflex but also laying the foundations for mental life in more complex organisms.[9] Antonio Damasio also came to this conclusion, with his proposal that consciousness consists of the brain making an image of the bodily processes underlying homeostasis.[10] More recently, Mark Solms has linked this process to the brainstem, arguing that this neural structure is the "hidden spring" of consciousness.[11] To all these thinkers, the life process is essential for under-standing experience, but it alone is not enough for consciousness itself. The problem is that all these extra steps move us back toward the hard problem, with the issue of how such neural mechanisms are related to subjective ex-perience. I suggest we simply resist making this move and reject the taboo of considering consciousness in living systems that do not possess nervous systems.

I should also take a moment to distinguish this proposal from an idea called biocentrism, proposed by Robert Lanza.[12] The standard picture of how the physical universe, life, and consciousness fit together proceeds from physics to life and mind. Lanza proposes a mind-bending idealist vision of existence that operates in the reverse order, with consciousness coming into existence through life, which in turn creates the fabric of the cosmos. Such a picture resonates with John Wheeler's idea of a participatory universe, in which the act of observation now collapses the wave function of the uni-verse to retroactively bring the big bang and the rest of the evolution of the cosmos into existence. I do not take this position and instead feel we are on firmest ground if we take biology to be grounded in physics rather than the other way around. While it may feel like we and our minds are at the center of the universe from our perspective, I believe this is only true from our perspective.

The belief in all life being conscious may be uncomfortable at first, but as it is presented here, it is entirely naturalistic and philosophically con-sistent. The same cannot be said for the belief that consciousness evolved, which invites the hard problem by assuming that it is possible for life-forms to exist that have no capacity for experience. Making this move inevitably relegates consciousness to the status of ineffective add-on. If we are to en-tertain the strange idea of experience existing throughout the entire living world, then we had better see whether scientific evidence indicates this as being plausible or not.

BEYOND THE BRAIN

The brain is often lauded as somehow fundamentally different from the rest of the body, yet we need not think of the brain in this way. It is an astounding organ, yet it is an organ nonetheless, made of biological cells like any other part of the body. We must see the brain in this light in order to understand that there is nothing about it that makes it uniquely capable of consciousness. What's more, the division of brain from body is one that only exists in our imagination. The nervous system is entirely continuous with the rest of the body, spreading throughout it from top to bottom. The magic of consciousness is to be found in biology, not in some abstraction of the brain that elevates it above the biological.

If you've ever undergone an operation that requires general anesthesia, then you'll know that these substances have a powerful effect on consciousness. The way in which anesthetics act on consciousness largely remained a mystery for a very long time and remains an area of ongoing research. You might reasonably assume that they must act on the brain in some way, presumably on the centers of the brain responsible for creating consciousness. If this is your assumption, then you may be surprised to learn that the same anesthetics that affect our consciousness also work on plants and bacteria, organisms with no nervous systems at all.[13] If anesthetics dampened consciousness by acting on the brain, this would make no sense because these nonanimal organisms do not have brains. We now know that anesthetics work at the cellular level, disrupting metabolism by acting on cellular respiration networks. This is clear evidence that the basis of consciousness is found in the life process and not in brains alone.

We cannot easily study consciousness in systems without brains that can't communicate with us about their inner world, but we can observe intelligent behavior for evidence of their mental lives. Intelligence and consciousness are not the same thing, yet intelligence is an integral part of our minds. For this reason, research on non-neural intelligence is a great place to look for support for the claim that minds can exist without brains.

The pioneering psychologist William James defined *intelligence* as the ability to reach a goal in a variety of ways.[14] Where does this ability come from? We often think of intelligence as a process of gathering data and submitting it to some form of analysis, a view that fits with the idea of it requiring a machine or brain with a sophisticated computational capacity. In reality all living things must problem solve in order to self-perpetuate and avoid being pulled apart by the second law. The relentless forge of selection is how intelligence came into existence. We saw with the theory of autopoiesis that living forms can be thought of as cognitive processes, indicating that we should not be surprised to find such capacities as memory and decision making throughout the living world.

Everything alive must solve problems in order to continue to exist. Biologist Michael Levin argues that we typically do not appreciate this because many creatures solve problems in spaces that might be unfamiliar to us. Humans often solve problems in physical space, as when procuring food, for example, but we problem solve in other spaces, too. Philosophers, mathematicians, and chess grand masters can sit motionless while navigating an abstract mental problem space of concepts in order to reach a goal. In fact, this is practically the standard, cliched image we have of intelligence. Levin argues that every cell in your body is also intelligently solving problems but in unfamiliar realms, such as DNA-transcription space. Seen in this way, there is no dividing line between the cognitive capacities of creatures like us and those of a single cell.

FROM BRAIN TO BODY

A caterpillar discovers a vast bounty of lavender plants with leaves that it can eat, but a strange-smelling chemical has been sprayed on it from a nearby farm. After being deterred at first, it eats and discovers that the chemical has no negative effect on it. It travels elsewhere, forms a chrysalis, and undergoes a transformation unlike anything you or I will ever experience. As part of its metamorphosis, its brain is liquefied, and a new brain is grown. A butterfly emerges and discovers plants with the same chemical on them. This time, undeterred, it drinks the nectar, feeling no threat from the chemical. Lab experiments have shown that experiences of this kind had by a caterpillar can be stored in memory and recalled after it undergoes its metamorphosis, despite the dissolving and regrowing of its brain.[15]

Information storage in the form of memory is a key aspect of intelligence. Could memory be stored in the body of an organism without the involvement of the brain? Michael Levin performed research with a flatworm called *planaria* to explore this possibility. This creature reproduces by dividing its body and effectively cloning itself. When a planarian is cut in half, both halves will regenerate the missing half of the body. How does the body know which parts to regenerate? Each cell contains the same genetic information, so Levin reasoned that the morphological information must be stored in the organization of the body, perhaps in the bioelectric properties of the somatic cells. To test whether this possibility was correct, his team removed both the brain-containing heads, as well as the tails, of a group of planarians, leaving just the middle section of the body. Such planarians would usually regrow both the head and the tail on the correct ends of this middle section. The researchers then mimicked the bioelectric signals of the head end on the tail end. The result? The torso regrew two heads, indicating that the body plan was indeed conveyed through bioelectric networks in the cells of the body.[16] What's more, when the planarians had their heads removed again, some of the remaining middle sections remembered the two-head body plan and regenerated both, even though the middle section that remained had no brain at all and nothing had changed in their DNA. It appears that it is indeed possible to store memories outside the brain.

Levin argues that this process of morphogenesis, the creation of bodily forms, is a kind of intelligence. The goal in this case is the creation of the correct body part, and the collective intelligence of the bodily cells must actively problem solve to navigate to this goal. When a rat navigates to a destination in a maze, it uses a network of electrically active cells in a brain region called the hippocampus to achieve this feat. When a salamander regenerates a limb, it uses an electrically active network of cells that rely on the same molecular mechanisms as the brain cells of the rat to "navigate" in morphological space to its destination of a fully regenerated limb. In the case of the salamander, these are cells of the body, and while their electrical dynamics are qualitatively different from those of neurons in the rat's brain, they still produce bioelectricity. Levin argues that the same principles that are at play in the brain are deployed by networks of bodily cells of this kind.

CELLULAR INTELLIGENCE

In the late seventeenth century, advances in microscopy made it possible for Antoni van Leeuwenhoek and Robert Hooke to discover the existence of microscopic organisms. It must have been incredible to behold this hith-er-to invisible universe of activity going on all around us for the first time, with minuscule animal-like creatures living lives of immense drama as they sought food and resisted succumbing to death. Over time, scientists would come to examine exactly what these little creatures were capable of.

In his 1906 book *Behavior of the Lower Organisms*, Herbert Spencer Jennings reported observations of his interactions with a single-celled organism called *Stentor roeseli*.[17] The trumpet-shaped stentor attaches to objects through a "hold fast," while they fan food into their mouths by creating a vortex with their hairlike cilia. Jennings squirted a dye into its oral cavity and observed that the stentor bent away from his dye-squirting pipette and rotated its cilia in the opposite direction to push the dye away from its mouth. It then contracted down to its hold fast and ultimately let go and swam away. Jennings's observations were replicated by researchers at Harvard Medical School, who found that these behaviors reflected a complex hierarchy of avoidance behaviors.[18] The decision to contract or detach was found to be 50/50, indicating that, rather than being a hard-wired response to its situation, it more resembled a choice being made on the part of the organism. Stentor has also been shown to be capable of learning, demonstrating the phenomenon of habituation, whereby it learns to stop producing a response in the presence of continuous stimulation.[19]

In the 1950s, psychologist Beatrice Gelber sought to test whether paramecia, a type of unicellular organism, are capable of learning.[20] She tested for associative learning, a prototypical form of memory that can be tested through Pavlovian or classical conditioning. This kind of memory was first demonstrated by Ivan Pavlov, who found that dogs could learn to salivate in response to the sound of a bell if it had been previously paired with the presentation of food. In a series of carefully controlled experiments, Gelber offered paramecia bacteria to eat, presented on the tip of a wire. After this, the paramecia learned to approach the wire even when no bacteria were present for them to eat, showing evidence of associative learning between the presentation of the wire and the presence of food. Gelber's research was dismissed at the time but has recently been revived. In 2021, researchers at Harvard reexamined her work and concluded, "Gelber was a remarkable scientist whose absence from the historical record testifies to the prevailing orthodoxy that single cells cannot learn."[21]

At some point in your life, it's very likely that you've walked past a slime mold. You may have observed these organisms living on tree bark, a fan-like yellow substance that most would not assume to be capable of anything particularly impressive. There are many kinds of slime mold, and this type is known as a plasmodial slime mold. Despite being capable of growing meters wide, they are acellular coenocytes, creatures that possess many nuclei but consist of only one cell membrane. Experiments have been performed with *Physarum polycephalum*, a particular species of this acellular plasmodial slime mold. This life-form survives by foraging for organisms, such as bacteria that live on decaying wood on the forest floor. This foraging behavior requires the slime mold to intelligently integrate trade-offs between multiple factors, including its hunger level, risks, and the quality of food available. By doing so, it manages to effectively allocate resources for growing in any particular direction.

Researchers who have sought to test the intelligence of this species have found that these organisms can navigate through mazes via optimal routes to reach food placed at a goal location.[22] The slime mold first explores the space and lays down a mucus as a kind of external memory trace so that it doesn't unnecessarily repeat journeys to where it has already been. After locating the food sources, it focuses its growth on the most efficient trajectory between the start location and the goal.

Many problem-solving tasks require groups of humans to collaborate in order to balance complex trade-offs. A classic example of this occurs in the design of infrastructure, such as rail networks, where cost, transport efficiency, and robustness must all be balanced in order to find an effective solution. Researchers in Tokyo sought to test whether this humble single-celled creature could solve such a problem, a task that usually takes teams of human engineers to solve.[23] They created a map of the greater Tokyo region and placed food in the locations of major cities. The slime mold created a network of links between the cities that not only resembled the actual network that human engineers had designed but also had comparable measures of cost, transport efficiency, and robustness. Such is the intelligence of this single cell that research is underway to put them into computer chips to boost the computational power of these devices.[24]

Another example of the genius of these creatures is in their ability to solve a problem that has haunted mathematics for more than a century: the traveling-salesman problem. If a salesman needs to travel to multiple locations once in a round trip, how should they determine the shortest route? It turns out that there is not an elegant mathematical solution to this problem that we know of, and computers will take exponentially long to run through the different possible solutions as the number of locations grows. Slime molds manage to solve this problem effectively, no matter how many locations are offered, without taking the dramatic hit to their processing time that occurs with computers.[25]

Another type of slime mold, called *Dictyosteliida*, or cellular slime molds, display a powerful collaborative-swarm intelligence. They live as individual unicellular organisms until food becomes scarce, at which time they form a swarm and team up to form a single multicellular slug. The slug then grows fruiting bodies that consist of a stalk with spores at the top. Only the individuals that become spores have the opportunity to be carried away on the wind and start the next generation. Those that form the stalk simply die and do not get to reproduce, leading some to call this behavior a kind of altruism.

FIXED MECHANISMS VERSUS EVOLUTIONARY PROCESS

When people try to make the case that unicellular organisms function by fixed reflexes alone, they often point to a simple behavior called chemotaxis. During chemotaxis, single-celled organisms, such as the bacteria *Escherichia coli* (E. coli), navigate toward concentrations of attractants, such as food, by alternating between "smooth runs," in which they move in a particular direction, and a "tumbling" behavior that allows them to reorient to a new random direction. By controlling the frequency at which tumbles are initiated, this little creature can successfully swim up chemical gradients to areas of high concentration of the attractive chemical. The tumbling frequency is adjusted by E. coli in sophisticated ways, depending on the chemical concentrations present.[26]

In order for a system to be an effective controller of an environmental variable (such as the concentration of a chemical) or of its own response to it (such as its tumbling frequency), that system must act as a model of the target variable. This idea is known as the good-regulator theorem. Even with the most basic behaviors of this kind, these comparatively simple organisms must process information to achieve their survival goals. According to the authors of this E. coli research, "It is extremely important to emphasize that the E. coli is a representational device that subserves the formation of internal representations of the world through networks of proteins."[27]

Scientists often come up with simplified models of behaviors in unicellular organisms and then make the mistake of thinking that their simple mechanistic model captures everything relevant that is going on in such creatures. In reality each cell is a universe of complex activity. Drew Berry of the Walter and Eliza Hall Institute of Medical Research has made animations of the activity that occurs inside cells that have received millions of views online, as has "The Inner Life of the Cell," an animation from Harvard University and the Howard Hughes Medical Institute in collaboration with XVIVO, a medical technology company. These animations depict this complex universe of interacting molecular components, yet even these impressive visualizations are still simplifications. The interconnectivity of subcellular processes creates a vast space of possibility for selection to produce impressive behavioral results when such organisms are faced with problems.

We ignore this vast network of complexity that makes flexible behavior possible for unicellular organisms when we think our simple models capture the full extent of what is occurring in such systems. Consider whether it would make sense for me to do the same when explaining your behavior. When you walk, your spine contains a simple circuit called a central pattern generator (CPG) that sends the electrical signals that make your leg muscles alternately contract in a coordinated manner. If I were to observe you walking down the road, I could describe this simple mechanical picture of how it is that your legs come to move and could then conclude that you are a simple automaton. In doing so, however, I would be ignoring your brain, with its near-limitless capacity to find flexible and innovative solutions to problems you might be faced with. We are doing the same thing when we dismiss "simple" organisms as hard-wired reflex machines. Even your capacity to innovate in countless situations could be reduced to a toy diagram of flexible synaptic connectivity in your brain, but this way of seeing misses the point of what life is. We are not fixed bundles of specific mechanisms; we are embodied evolutionary processes. We are an inherently open-ended, adaptive phenomenon, not a machine. This is as true for bacteria as it is for *Homo sapiens*.

PLANT INTELLIGENCE

One weekend, while working at UC Berkeley, I hiked into the hills of Claremont Canyon Regional Preserve and sat on the ground, overlooking the city to the San Francisco Bay. As I looked out over the streets, my attention landed on a large tree on one of the roads. I suddenly saw it as another living thing, a sentinel-like being, a silent inhabitant of its little street. In that moment, it had gone from being seen as an irrelevant background object to being recognized as a relative. With this new way of seeing, I suddenly noticed another tree and another and yet another. I slowly saw that the city was swarmed with these living things, these unassuming organisms that quietly carried out the activities of life while the bustle of a city passed by around them. They looked as if they were lying in wait, ready to retake the land the moment humans left. I let my attention widen, and suddenly I perceived the whole city as a grotesque concrete scar that had been dropped on the glorious living world beneath it. In the moment, I felt reassured by a sense of inevitability that, given enough time, the trees and their green brethren stifled beneath the concrete and asphalt would inevitably come to reclaim this patch of earth, hopefully with us living alongside them rather than as a result of the demise of our species.

Over the past several decades, research into the mental capacities of plants has been coming on in leaps and bounds. The research community has been building a convincing body of evidence that plants possess cognitive capacities that are closer to our own than many had ever thought. Intelligence is not experience, but if plants can be shown to demonstrate memory and problem-solving abilities, then it supports the idea that we may share other aspects of the mind, such as consciousness.

We did not evolve to pay much attention to plants, beyond the amount required to consume some of them and to avoid poisonous ones. A vine is unlikely to chase you down and eat you, so it makes sense to reserve your brain processing for fast-moving saber-tooth tigers and other dangerous animals. This creates a phenomenon in humans that has been termed plant blindness, in which plants are seen less as living agents and more as irrelevant scenery that retreats into the background.[28] If you take a time-lapse film of certain plants, however, their sped-up movements look strikingly like the behavior of a living agent, as that is indeed what they are.

I remember when I first learned that plants are alive, in biology class as a child. Perhaps it was because I was at a Catholic school, or perhaps it was wider cultural conditioning, but I remember being shocked to learn that plants were living things. Trees turned out to not be the inanimate objects that I had thought they were and were revealed to be our distant living cousins. The realization that I couldn't trust my commonsense assumptions and instead had to use science to ascertain the truth never left me. This is an important lesson to keep in mind when exploring the mental lives of organisms unlike ourselves. We must rely on science and philosophical rigor and not our biased intuitions in order to find the truth.

Rather than being simply reflexive, plants exhibit intelligent, goal-directed, flexible behavior that shows evidence of both decision making and learning. Latzel and Münzbergová managed to "teach" wild strawberries (*Fragaria vesca*) to associate light intensity with the availability of nutrients in the soil, showing an ability to learn through classical conditioning.[29] Other studies have found the pole-grabbing behavior of beans to be goal directed, flexible, anticipatory, and affected by learning.[30] Further evidence for anticipatory behavior, as well as for the presence of internal models of the outside world in plants, comes from how these organisms orient toward sunlight. It is sometimes suggested that plants simply respond reflexively to the sun's light, but some plants show tracking of the sun's position even when it is cloudy and can predict where and when the sun will rise.[31] This provides evidence that these plants contain some kind of internal model of the sun's position over time.

Plants even appear to show a form of social intelligence, demonstrating the ability to recognize relatives. American sea rocket (*Cakile edentula*) plants that are grown in a pot with strangers will grow more aggressively, producing a larger root mass than when grown among their own species.[32] Communication is also common in the plant world. Some plants release volatile organic compounds (VOCs) to signal to other plants. The smell of freshly cut grass is such a communication act, a signal to the neighboring blades of grass to rally their chemical defenses against an imminent threat.

Trees have been found to exhibit prosocial behavior, too, including warning their neighbors of incoming damage from insects via VOCs. In *The Hidden Life of Trees*, Peter Wohlleben reports on research that has found that trees suckle their young by providing them with sugars until they can make their own from photosynthesis.[33] The same process happens with sick trees being kept alive by their healthy neighbors. Wohlleben even discovered a beech tree stump that had been kept alive in this way for five hundred years, despite it having no leaves with which to produce its own sugars. Such transfers happen through what researcher Suzanne Simard has called the wood wide web, a network of fungi that have evolved to collaborate with the trees.[34] It has been argued that fungi, too, possess the capacity for decision making, as evidenced by individual variation in their growth patterns.[35]

With each new study, there is increasing evidence for creatures without nervous systems possessing the intelligent problem-solving capacities that we should presume they must, given our earlier exploration of the thermodynamics of life. These findings are in keeping with the ability to problem solve being a necessary feature of life, suggesting that the idea that these organisms can survive by reflex alone is a misconception. The idea that any organism can afford to sit back in a cushy niche, with no need to problem solve in order to continue to exist, is a myth.

EMBODIED BELIEFS WITHOUT NERVOUS SYSTEMS

Do plants possess the kinds of physical structures that would be necessary for mediating conscious experience? Plant physiology is not something entirely different than brain physiology. Plants use many of the chemicals that function as neurotransmitters in the brain, including the two main neurotransmitters glutamate and GABA, as well as neuromodulators, such as dopamine, serotonin, and melatonin. These chemicals are associated with certain functions in humans, but it is not the chemicals themselves that carry the ability to produce these effects. Neurotransmitters activate specific circuits in the human brain, but it is the circuits that are responsible for the function, not the chemicals themselves. For example, in humans, melatonin can trigger sleep, but all the melatonin molecule itself does is activate certain receptors in the sleep circuitry of the brain. It is a lock-and-key mechanism that can be repurposed for many other functions in different organisms. The molecule GABA was originally found in plants, but GABA was later found to be the main inhibitory neurotransmitter of the brain, serving to dampen electrical activity. In plants, though, GABA has been found to play a role in defense against insect damage.[36]

Such similarities gave rise to the controversial term *plant neurobiology* for this field of study.[37] The controversy stems from the fact that, while plants do show biochemical overlap with the realm of neurochemistry, plants do not possess neurons, from which the term *neurobiology* is derived. They do engage in bioelectrical signaling, however, leading plant neurobiologists to advocate for the legitimacy of this term. The main excitatory neurotransmitter in the brain is glutamate, and it produces changes in electrical activity through the release of calcium. Wounded plants have also been found to use glutamate to trigger the release of calcium.[38] What's more, the genes involved in glutamate signaling are similar in animals and plants.[39]

Some researchers speak of a phytonervous system, which is argued to consist of the vascular bundles of the plant transport system. One researcher, Paco Calvo, teamed up with Karl Friston to explore how predictive processing of the kind described in the free-energy principle (FEP) could be mediated through plant physiology.[40] Calvo points out that electrical signals propagate through the vascular bundles of plants throughout the entire organism in both directions. He argues that this bidirectional signaling could implement the transmission of predictions in one direction and prediction errors in the other direction, so that the surprise-minimizing algorithm of the FEP could be implemented.

Could single-celled organisms implement predictive processing? František Baluška, Arthur Reber, and William Miller Jr. have proposed that single-celled organisms contain what they call nanobrains, structures capable of sophisticated information processing.[41] The cell consists of a plasma membrane that divides it from its environment but through which it also senses the outside world. Movement is achieved through changes in the cytoskeleton, a dynamic network of filaments. These researchers suggest that the plasma membrane of cells functions as the first nanobrain that they argue could mediate consciousness. A second nanobrain is suggested to comprise parts of the cytoskeleton and associated structures. We can imagine these two nanobrains as two information-processing structures: the first dedicated to sensation, and the second, to action.

For any living organism, beliefs about the world beyond should exist in this loop from sensation to action. In the case of single-celled organisms, the processes that instantiate these probability distributions regarding patterns in the outside world should be found between the sensitive plasma membrane and the cytoskeleton that implements movement. The sensorimotor loop of such creatures begins with receptors on the plasma membrane being activated, often by chemical or mechanical activity. These activated receptors then initiate intracellular signaling cascades, in which complex networks of molecular interactions take place. This biochemical process results in cytoskeletal regulator proteins being activated, proteins that directly puppeteer the cytoskeleton, leading to changes in the organization of cytoskeletal filaments: that is, movement.

These intracellular molecular-signaling cascades are one process through which beliefs could exist, mediating the mapping from sensation to action. This would be analogous to beliefs existing in the activity of neurons in the brain. In the nervous system, it has been suggested that these beliefs could also exist in structural patterns of connectivity. Perhaps the physical organization of the filaments themselves could serve a parallel function in single-celled organisms. These claims are highly speculative at this point, as research on the dynamics of predictive processing in single-celled organisms is not being conducted, to my knowledge. The key takeaway here is that all living systems possess the requisite loops from sensation to action within which predictive processing could be implemented, permitting consciousness to exist in all living things. We are not conscious because we have brains; we are conscious because we are alive, and there is no reason to not generalize this link between life and experience to all other living things.

FOURTEEN

CLOSING THE GAP

It is not necessary to ask whether soul and body are one, just as it is not necessary to ask whether the wax and its shape are one.—ARISTOTLE[1]

Shifting from a human-centered, brain-centered, or even nervous system–centered view of consciousness to a life-centered view allows us to see many aspects of the mind-body problem in a new light. Here I flesh out the precise details of this new understanding of experience and use this perspective to illuminate multiple mysteries of the mind. How exactly can we connect matter to mind? Can machines even be conscious? What happens to awareness when we sleep? How are we to understand spiritual states of consciousness?

EXPLAINING CONSCIOUSNESS

It is easy to get the impression that science progresses by proving theories to be correct. This is not typically how science works, however. The acceptance of any explanation usually happens due to it both fitting with the data and because it offers an account of the phenomenon in question that is intuitively satisfying. Intuitive appeal is set aside in rare cases where the fit to the data is exceptional, as in the case of quantum mechanics. What we are looking for in a theory of consciousness is something that both fits the experimental data and convincingly answers the question of why experience exists.

We have many widely accepted intuitive explanations for different natural phenomena, such as why water flows at different temperatures, but not for consciousness. In the case of water, if you asked me why ice doesn't flow and warm water does, I could respond that the strength of the bonds between the water molecules changes with temperature. At warmer temperatures, the bonds are weak, allowing the molecules to tumble over each other. As the water cools to freezing, the bonds become strong, preventing the tumbling. These dynamics at the molecular scale provide an intuitive account that makes sense of the properties of water at a larger scale. I suspect most people consider this to be a satisfactory explanation for this phenomenon.

If you asked me why you experience the sights, sounds, and smells of the world around you and I told you it was because your brain activity is organized into waves that peak roughly forty times a second, you most likely would not consider the case closed. This was an early scientific theory of consciousness, put forward by Christof Koch and Francis Crick.[2] Unlike the water explanation, it doesn't give insight into why consciousness should exist, given these particular dynamics in the brain. Why should brain oscillations give rise to the smell of coffee or freshly baked bread? Why forty times a second and not fifty? Such explanations do not solve the mystery of consciousness.

Here is an answer to the question of why consciousness exists in systems like us and not in rocks, for example, that I personally find intuitively satisfying. A rock is not conscious because it is not set apart from the world around it in any meaningful sense—it is part of the thermodynamic slurry of existence that will be worn away into greater and greater disorder. Today, it appears as a rock; tomorrow, it is a million grains of sand. Unlike a living thing, it never attempts to resist this inevitable descent into disorder. What we call "the rock" is just the location of certain atoms bound together at a particular place and time. Unlike life, it shows no behaviors that meaningfully differentiate it from what we would consider its surrounding environment. As a result, there isn't even a "thing" there that we can inquire about the consciousness of (as explored in chapter 2). You, on the other hand, are an island of order that resists the second law. To actively maintain this order, you must infer the nature of the world around you and of yourself within it. If you didn't, you wouldn't be able to exist over time. You would be sucked into the thermodynamic chaos of wider existence, like the rock. This is what happens at death, when we presume consciousness ceases.

This gives us an explanation of why you—and living things like you—are associated with the phenomenon that we call consciousness and why the most fundamental of our conscious feelings are tied up with our survival. Living things are self-maintaining islands of order that manage to anticipate and avoid the destructive influences of the outside world. Consciousness is an interface that such living systems use to navigate a world that is forever out of reach, the existence and structure of which we must infer. While not truly separate from their environment, living things attempt to impose a boundary to prevent destructive forces from disturbing their order. It is in this interaction with the world that a living thing constructs a vision of the reality that it inhabits.

This is an account that connects the physics of life to the dynamics of conscious experience, resolving the mind-body problem. In this view, conscious experience exists as a systemic process, a dynamic interplay between an organism and its environment. The human brain provides a hugely complex elaboration on this process, but it is not the source of consciousness in the first place—it is the evolutionary inheritor of it.

SOLVING THE HARD PROBLEM

Intuitive appeal is one thing, but philosophical consistency and rigor is quite another. Does the explanation for consciousness offered here really close the gap between matter and mind? How can we jump from the objective operation of biological processes to subjective conscious states? I start this book with an exploration of the hard problem of consciousness, the idea that no quantitative mechanism can explain qualitative experience. The way the hard problem was originally formulated rests on the "philosophical zombie" thought experiment, which asks us to imagine creatures like us that perform the same functions as us but without conscious experience.[3] Chalmers used the fact that we can imagine such a being to argue that consciousness must be something different from physical processes. If one believes consciousness to be limited to human beings, then the world is full of such philosophical zombies. If one believes that insects are not conscious, then they have eyes but do not see. They process visual information but do not have visual experience, a very odd state of affairs. In the view presented here, however, this is simply not conceivable. With this new understanding, it is not possible to imagine that a philosophical zombie could exist, as to persist over time, it would need to engage in the belief-formation process that I have argued is equivalent to consciousness. We can imagine an airplane flying backward, but once we have an understanding of the aerodynamics that make flight possible, the possibility of such a thing actually occurring in the real world becomes inconceivable. The same is true with life and consciousness. If life and consciousness go hand in hand, then it would feel like something to be any living thing as it navigates in the world. With this understanding, the artificial delineation between conscious and unconscious living things dissolves and with it the hard problem.

If we move beyond the specific details of how Chalmers framed the hard problem, then we can still ask how one can connect the objective description of the behavior of a system to the subjective world of experience. How can describing objective quantitative processes account for subjective qualitative feeling? An explanation of consciousness will not collapse these two descriptions into one, but a successful theory should be able to capture consciousness using both kinds of description in a way where it seems reasonable that they really are describing the same phenomenon. The claim here is that (1) the subjective description of consciousness as a framework of beliefs in the qualitative character of the world and (2) the objective description of how and why such a framework of beliefs exists in living things are both reasonable descriptions of consciousness. We cannot prove that these descriptions are of the same phenomenon, much as we can't directly prove that the loose bonding between H_2O molecules is a satisfactory explanation for the wetness of water. This equivalence feels right to us, in part because it fits with the data of how water changes with temperature, and so we accept it. We could argue that there is a hard problem of water, consisting of the unbridgeable chasm between these two descriptions, but instead we tend to be happy that these are two descriptions of the same thing. It would be pedantic to focus on the fact that you can't prove they both refer to the same phenomenon. A successful theory of consciousness should also possess this additional kind of intuitive appeal.

It is also important to emphasize that the explanation offered here is not one that is framed in terms of a reductive mechanism but instead in terms of properties that emerge with certain modes of organization, specifically the property of being conscious emerging with the mode of organization we call life. The gap between mind and matter arose with Descartes and his vision of a world of two substances: the objectively observable physical world and the private inner world of experience. This metaphysical gap is resolved here through a perspective that is based on Spinoza's neutral monism (a statement that would have undoubtedly disappointed my eleventh-great-grandfather whom I mentioned in chapter 4). All of reality is a relational process. When we study the publicly observable aspects of reality, we call it the physical, but there are also relational dynamics that are not publicly observable, ones that exist through the emergent organization of living things, relational dynamics that we call consciousness. Consciousness is a complex form of relationality that exists in the systemic organization of an organism and the environment it is embedded in. This means that consciousness is a kind of contact, a dynamic that exists at the intersection of the organism and the environment, a loop between the two that involves both. In this view consciousness does depend on matter but in the way that a complex pattern can be built from simpler patterns. The continuity between matter and mind here rests on the idea that only relations exist. If the stuff of physics and consciousness are both forms of relations, then there is no gap to be crossed between matter and mind.

Why is it necessary to conclude that all living things are conscious? Why couldn't consciousness be confined to more complicated creatures like ourselves? In order to understand what consciousness is, we must appreciate that it is not a *thing* but a *process*. Consciousness is not an optional extra but a fundamental aspect of living. When you burn your hand on a stove, the pain you feel and the loop of sensing and responding are the same thing, experienced from the inside and observable from the outside. There is no extra "thing" that is the conscious experience. If you assume that some living things show such behavior but possess no potential for experience, then you are conceptualizing consciousness as something additional. Dissociating what are two sides of the same coin is what got the field of consciousness science confused in the first place. This is a core mistake we must correct.

COULD CONSCIOUSNESS HAVE EMERGED WITH BRAINS?

If the whole of the organism interacting with its environment is synonymous with consciousness, then why do only certain aspects of brain activity correlate with conscious experience in humans? Consciousness comes into existence with living bodies, and in us the brain is a part of the body that is involved in structuring the contents of consciousness. The brain evolved as a sophisticated organ for predictive processing, with certain parts of the system correlating more with the contents of consciousness than others. I explained earlier that the information content of primary visual cortex, V1, appears to correlate less with conscious experience than that of higher cortical areas. Similarly, while the retina is necessary to consciously perceive objects in front of you, there isn't a strict correlation between its activity and the nuances of what is perceived: for example, as shown in the paradigm of binocular rivalry in which the images on the retina are kept static while the conscious experience flips between two different images. Why does this pattern of correlation between certain brain areas and conscious content exist?

According to the free-energy principle (FEP), living things both sense the world and respond to it, and in between, the internal states of the organism hold beliefs about the hidden causes behind the sensory input it receives. It is the belief formation in the middle of this process that structures the contents of consciousness, so we should expect the neural structures in the brain that instantiate these beliefs to correlate with the contents of experience. There is nothing magical about these brain areas; they are just the nodes in the network that co-vary with the contents the most due to the functional role they play. This is what gives rise to the neural correlates of consciousness. I return to this process in more detail below when accounting for the precise contents of consciousness.

Let us consider for a second what would have to be the case for experience to come into existence with brains. In order for a genuinely novel phenomenon like consciousness to develop, some emergent mode of organization would need to have arisen to underpin it. This extra level of emergence would have needed to have taken place between single-celled organisms and us. Candidate transition points could include multicellularity, the development of nervous systems, or the evolution of a certain function only performed by particular kinds of nervous systems. These are interesting phenomena, but none of them provide the kind of discontinuity we see between nonlife and life, and such a discontinuity is necessary for genuinely novel emergent phenomena to come into existence. Instead, they represent the gradual, continuous morphing of species throughout evolutionary history. We do not have separate fields of scientific study for different kinds of living things. Inorganic chemistry and biology study different levels of organization due to the fact that biology is a genuinely emergent phenomenon, but we do not see such divisions within biology. You may argue that plant biology is a different field from animal biology, but it's important to appreciate that there are specializations within the same overall field of study.

If we consider consciousness to be a biological phenomenon, then we must place its origins with the emergence of life. To do so raises no philosophical problems, as far as I can see, while placing it at any point in evolution after the emergence of life invites the hard problem once again. If living things could function just fine without consciousness before a certain point, then what function could it possibly be performing? We can simply avoid this intractable philosophical issue from arising by taking the stance of biopsychism. The emergence of life is a transition that contains all the necessary processes to underpin the arising of meaning-filled experience.

A common way to argue that organisms without nervous systems are not minded is to suggest that, while we *have* models of the world stored in our brains, these creatures merely *are* models of their world. In the case of the giraffe, its bodily form is a model of its environment: The length of its neck correlates with the height of the leaves that it must eat, for example. Evolution molded the structure of this organism to fit the structure of the environment, much as the pattern of notches on a key fits the structure inside the lock that it opens. All bodies *are* models of the niche that their species evolved in. These bodies must also possess inferential survival dynamics that mediate between sensation and action in order for them to persist over time, and it is the models of the world that are instantiated in these dynamics that all living systems *have*. It is these models that I propose are equivalent to consciousness. The difference between being a model and having a model is the difference between the form and the dynamics of a system.

To disprove the idea that all living things have models of this kind, we would need to identify an organism that is truly only a body with genetically hard-wired reflexes, one that does not possess its own adaptive capacity to respond to its environment. We cannot definitively say that there is an organism alive that lacks this capacity. It would seem to me that such a system wouldn't even qualify as being alive. For such an organism to exist, the selection algorithm would have had to put on the brakes at the level of bodies. Selection-based adaptive procedures, such as learning, would need to have not be selected for, something that would appear odd given the requisite biochemical machinery existing in all life. I see no way for organisms lacking such adaptivity to resist the second law and do not believe they exist.

Might the lights have gradually come on throughout evolution with the development of nervous systems rather than suddenly with the emergence of life? Philosopher Peter Godfrey-Smith, in his book *Metazoa*, argues that consciousness is a gradual phenomenon of this kind.[4] The same proposal has been made by a group of researchers who suggested a metaphysical framework called Markovian monism, which links consciousness to the free-energy framework but claims it emerged gradually from the first unconscious living things to us.[5] Karl Friston has suggested that the magic ingredient that makes something conscious might be the counterfactual and temporal depth of the beliefs it holds, which permits a system to model itself as the subject of experience.[6]

It is one thing to simulate the world as it is, but we can also imagine it to be ways that it is not; this is what is meant here by the term *counterfactual*. Counterfactual depth refers to the extent to which a system can do this. Temporal depth refers to how far in time the system can simulate what might be occurring in the world. When a system's beliefs are sufficiently deep, a system can model the consequences of its own actions and imagine multiple ways in which it might navigate the world. According to Friston, it is this capacity of self-modeling as a result of such depth that makes a system conscious. Tomas Metzinger has pointed out, however, that the recognition of pure consciousness, or what I have called formless awareness, is associated with a collapsing of temporal depth into the feeling of a seemingly eternal or timeless now and also does not depend on a self.[7] Friston's model accounts for aspects of the contents of consciousness, such as the sense of self, but I do not believe it accounts for the emergence of consciousness itself.

THE ORIGINS OF AWARENESS

A major problem for the idea that consciousness emerged gradually through the complexification of nervous systems or by any other mechanism is the nondual nature of the formless awareness that lies at the heart of consciousness. The brain is an incredible organ that surely elevates conscious content to an astonishingly rich level when compared with the simplest organisms, but this is a difference in degree, not kind. It is the contents of consciousness that become enriched by the complex processing of the nervous system, not the fact of awareness that lies at its core. The experience of a unicellular organism may be immeasurably different from our own, but where there is any experience at all, the all-or-nothing phenomenon of nondual formless awareness must be present.

Awareness is nondual in the sense that it is unaffected by whether the contents arise from what we consider the subject or what we consider the world; it is a pristine, empty space that is unaffected by the contents that arise in it. While the contents can change gradually, awareness itself cannot. As the space in which contents arise, it is either present, or it is absent. It is not possible for a space to exist a little bit—either it does, or it doesn't. Given its total lack of properties beyond the capacity for awareness, the idea that this empty experiential background could be the same for all creatures is compelling.

The idea of awareness gradually coming into existence has some appeal, in part because it allows us to avoid being specific about precisely where and why it emerges. It might seem like a strange idea to imagine one's own perspective suddenly popping into existence at a single moment, as would be the case with the nongradual account of awareness described here, but we assume the transition in the opposite direction happens when we die. The contents of consciousness may fade gradually, but the recognition of the nongraded nature of awareness commits us to the idea of a binary transition happening at the moment of death. How can we account for such a sudden switch? Living agents show end-directed, teleological behavior (i.e., behaviors that can be explained in relation to the purpose they serve), in that they are systems that are organized around the goal of persisting. Rather than passively dissolving over time like nonliving things, it is as if living things shift into a reverse direction and organize themselves around their individual futures. We can conceptualize the difference between life and death as the net movement in either the teleological direction of being organized around the goal of survival and successfully resisting the second law or moving in the other direction of passively melding with everything else in existence.

In thinking about life and death in terms of a net positive or negative direction of this kind, we can see a natural point where a threshold could occur, where on one side a system is not conscious, and on the other it is. This is the zero point at which there is no net movement in either direction. In this picture the space of awareness sprung into existence for the first living system when it began functioning as a holistic emergent agent (net positive movement in the direction of self-persistence), and this bubble of experience pops at the moment of death (net negative movement heading toward entropic decay).

If awareness is present throughout our entire lives, then it must be present even while in deep, dreamless sleep. Is it really possible that formless aware-ness can be present without content and without us remembering such moments? Highly trained meditators report that this awareness persists at all times, even throughout sleep.[8] We typically assume that consciousness disappears every night when we go to bed. It is possible, however, that it is only the conscious content that diminishes, including the content we call the self. If this were the case, then it would be no surprise that the cognitive machinery that weaves together the feeling of being a self that has expe-riences and arranges such experiences into a narrative so that they can be recalled later is not functioning at such times. This may be what gives the impression of a total absence of consciousness during dreamless sleep, even though formless awareness continues to be present. The space of awareness indeed appears to be ready for you upon waking. We might assume it comes online with the content, but perhaps it was there, operating in the back-ground, ready for you to start your day.

WHY ARE THERE NEURAL CORRELATES OF HUMAN CONSCIOUSNESS?

There must be an infinite number of degrees of consciousness, following the degrees of complication and aggregation of the primordial mind-dust.
—WILLIAM JAMES[9]

So it is possible that formless awareness persists during sleep. Once you do start your day, however, we experience a rich array of conscious content. If I were to scan your brain while you were having these conscious experiences,

I would see correlations between what you experienced and certain patterns of brain activity. What are we to make of this connection in light of the proposal made in this book? We have seen that the correlates of consciousness are to be found in the cortex and that oscillations at around forty hertz are linked to conscious content. We are now well placed to explain why this should be the case.

In suggesting that consciousness is synonymous with the embodied predictive processing performed by an organism, how are we to account for the specific contents of consciousness? In the predictive-processing framework, we typically think of there being a "winning" hypothesis in any situation, the one that is best suited to the specific circumstance. This corresponds to the hypothesis with the largest posterior probability in Bayesian terminology, the one that is most likely after combining the evidence with one's prior assumptions. How is this hypothesis supposed to kindle an experience, while others do not? If we think of such beliefs as existing in a brain that is disconnected from the outside world, then this does seem like a difficult problem to answer. The losing hypotheses are mere unconscious brain activity, while the winning one somehow brings an ephemeral qualitative experience into existence. If we adopt the embodied relational view offered here, however, this process need not be seen as a magical or mysterious process.

Consider a simple organism that moves away from something painful. The hypothesis or belief that the stimulus is painful exists in the sensory motor loop involving both the outside world and the organism that leads to this creature responding by moving away from whatever is producing the pain. The losing hypothesis that the stimulus is not painful doesn't come into the equation, as that behavioral loop is not recruited. Conscious content is selected for based on which response is most conducive to survival. The resulting embodied dynamics of the organism and the experience are the same thing.

How does such a selection process play out in the human brain? Earlier I mentioned that Gerald Edelman suggested a theory of brain function called neural Darwinism, in which populations of neural activity are selected for based on how well they serve the organism's quest for survival.[10] This process may underlie a rapid rewiring of patterns from sensation to action that instantiates the relational dynamics that correspond to conscious content. Crucially, the selection of the appropriate neural populations is implemented through a process called reentry, in which the selected patterns are amplified through feedback loops of neural activity. The circuit that mediates this feedback loop involves mutual connections between the thalamus to cortex that oscillate at around forty hertz, producing the very brain activity that has consistently been found to correlate with the contents of consciousness. Now we can see why brain activity at forty hertz is a neural correlate of consciousness, while appreciating that it does not generate consciousness.[11] Experience comes into existence with embodied living things, and this selection-based brain process modulates it in humans, shaping the content of our consciousness.

It is one thing to explain the emergence of consciousness—that is, why it exists at all—and another to account for its precise contents. The predictive-processing theories of consciousness mentioned earlier are all neurocentric theories of consciousness, so they fall into the trap of the hard problem. As a result, they do not explain the emergence of this phenomenon. They do, however, provide excellent accounts of how we can understand the contents of consciousness that are associated with the human brain. This book is focused on the question of how consciousness came into existence, whereas these models take us into the details of human experience.

In chapter 8, I discussed the dispute about whether the process of homeostatic regulation performed by the brainstem or the higher-level functions performed by the cortex should be thought of as underpinning consciousness in the human brain. If we see consciousness as not being generated by certain brain areas but understand it to exist as a systemic phenomenon, then we can resolve the disagreement between these positions. We can think of both the internal interoceptive regulation of bodily processes and the perceptual navigation of one's external environment as two necessary ways of achieving the broader goal of surviving over time that structures all living things. With this conception, both the bodily regulation performed by the brainstem and the perception of the world and oneself in it that is mediated through the cortex are equally part of the same overarching function of survival. We should think of an individual human as a holistic organism in the same way as a cell. We are made up of smaller cellular components, but the principles of survival that drive our overall organization are the same. As a system, the brain functions to facilitate our survival, as complex molecular pathways do for single-celled organisms.

The complexity of how the human brain elaborates the contents of consciousness to produce our rich lived experience will surely provide fertile ground for scientific study for many years to come, but the issue of why these contents are experienced need not remain a mystery, if we appreciate our consciousness and that of a bacterium to be the same organism-level relational process of holding beliefs about existence.

Using the idea of causality through constraint explored in chapter 10, we can now see how a circular causality can exist between consciousness and the neurons that make up the brain.[12] Recall our examination of the ways in which parts and wholes influence each other in the example of a flock of birds. Consider how it is that individual birds contribute to the behavior of an entire flock, while the organization of the flock shapes the trajectories of the individual birds. Here, the ways the birds initially interact can be understood as a set of enabling constraints that establish the initial overall pattern, while the organization of the flock can be considered a governing constraint that shapes the overall evolution of the whole system. Neurons are like individual birds in a flock, with overall brain activity being like the holistic flock formation, a pattern in the whole that can guide the behavior of the parts. This picture accounts for how it is that our seemingly immaterial experiences can produce changes in the activity of our physical brains and how these mental forms can produce tangible physical changes in the movement of our bodies and in the world beyond.

Imagine seeing a photograph of a friend and then deciding to call them as a result. It appears that the pattern of light falling on your retina plays a causal role in generating your conscious experience of perceiving your friend's face and that this conscious experience subsequently causes your muscles to move so that you pick up the phone and call them. What is occurring here? To begin with, the image lands on the retina and is transformed into electrical activity that is then passed on to the neurons of the brain. These signals can be thought of as the enabling constraints that set up the particular overall dynamic at a larger scale within the brain, at the scale of particular large populations of neurons. These are the neural populations that encode big-picture beliefs, such as the belief that you are seeing your friend's face. This mode of organization then evolves to an associated belief state, the belief that you should call your friend. This large-scale brain activity acts as a governing constraint mode, sculpting the activity of specific neurons so that they send the appropriate signals to your muscles. As a result, the conscious experience of seeing your friend's face results in your physical muscles moving to implement the planned action of making the call. In us humans, the neurons of the brain influence the conscious content that arises, and this content then shapes future brain activity. Consciousness is no mere epiphenomenon, experience is capable of producing tangible changes in the world.

THINGS THAT ARE CONSCIOUS

As a conscious organism, I have a particular perspective on the world, whereas you have a different perspective. What makes you and me two separate observers of existence, capable of having our own vantage points? Integrated information theory (IIT) attempts to address just this issue and how it relates to consciousness. The creator of IIT, Giulio Tononi, developed a measure of how integrated an information-processing system is, a quantity he calls φ (phi).[13] This approach is used to measure where irreducible wholes exist, systems that cannot be decomposed into their parts without damaging their organizational structure. Tononi speculates that this property of self-integration, or holism, that is measured by phi is what consciousness is. Myself and others, including Shamil Chandaria (see chapter 3) and Adam Safron, creator of the integrated world modeling theory (IWMT) of consciousness, have suggested that IIT captures a necessary prerequisite for a system to be a conscious observer, in that any system capable of performing predictive processing first must exist as a systemic whole of the kind described in the IIT framework but that this integration itself isn't what makes the system conscious.[14] Alicia Juarrero has suggested something similar, arguing that the holistic constraint regimes that govern emergent self-organized systems create the possibility for such systems to take on a perspective of their own.[15]

This conception of irreducible wholes gives us a way of thinking about which systems can be said to be conscious and which can't. Self-organizing systems like life show this kind of organization, whereas rivers do not. A river may look like a "thing" to us, but this is merely a conceptual overlay that we put on top of the messy continuous reality, where there is no true divide between falling rain, the water coursing across the riverbed, and the waves of the ocean. The whirlpool that arises when you drain your bath temporarily emerges as an organized whole, but without the predictive self-maintaining dynamics that we see in life, there is no reason to believe it to be conscious.

The process of evolution can be modeled as a process of inference, with organisms being seen as hypotheses about what forms best fit a given niche. Given that consciousness, too, can be seen as a process of inference, should we think of evolution as having its own exotic form of consciousness? Because it does not function as an integrated whole as individual organisms do, I do not believe we should think of it as being capable of supporting a perspective on the world. If a group of humans were to colonize another planet and were to never return to earth, while the rest of us stayed behind, then evolution could act entirely independently on both groups. The process is not an irreducible whole that can be disrupted through this kind of division. If half of your body were to suddenly be transported to Mars, however, you would not fare well. It is this integrated, holistic organization that makes it possible for you to be a conscious observer, while evolution cannot. In the case of evolution, there is no whole to inquire into the consciousness of. It is more like a river than an organism.

WHAT GIVES RISE TO CONSCIOUS CONTENT?

Life is a luminous halo, a semi-transparent envelope surrounding us from the beginning of consciousness to the end.—VIRGINIA WOOLF[16]

What is it about life that makes it different from nonliving things? Why not ascribe consciousness to other integrated relational systems that are not alive? For example, if you attach two clocks with swinging pendulums to a shelf, the pendulums will come to synchronize through vibrations that pass along the shelf.[17] The FEP, which forms part of the exploration of the dynamics of life and the connection to consciousness offered here, can also be used to model this system. Earlier I showed that something carries information about another thing if there is a correlation in their states, such that observing the state of one reduces our uncertainty about the state of the other. In this case, each pendulum carries information about the other when they are synchronized, much as Bayesian beliefs inside an organism come to carry information about the state of the world.

Through the lens of the FEP, the shelf acts like the boundary between a given pendulum and the "world," the other pendulum in this case.[18] The vibrations that pass along the shelf function both as sensation and action in this interpretation, detecting the movement of the other pendulum and responding in kind to create the synchronization. Should we say that each pendulum is conscious, given that they can be modeled in a similar way to life?

In the case of the pendulums, as with all physical phenomena, a relational phenomenon is occurring, but we should not think of this as the kind of complex relationality that constitutes consciousness. What is the key difference? Living things can navigate in the world in order to achieve the goal of maintaining their order, something the pendulum cannot do. This capacity of life is sometimes referred to as autonomy, although I feel uncomfortable with the characterization of organisms as truly autonomous, as it suggests some degree of genuine separation from the causal influences of the rest of existence. We can, however, characterize this phenomenon as agency.

An agent is a system that can set goals and can make choices in order to achieve those goals. We should not think of this capacity for choice making as being based on a kind of free will that would allow the agent to transcend the influence of physics and causality, however. Despite not possessing this Godlike ability to be the sole cause of its actions, agents do have the ability to act as the proximate cause of their own behavior. If I am not being coerced to do so, then my choosing tea over coffee is a decision that arises from within myself. The full picture of the causality behind that choice not only involves factors that were not in my control, such as being raised in the tea-drinking United Kingdom, but also the entire evolution of the universe and life on earth that led to my physiology as a fluid-consuming primate. Despite this bigger picture operating in the background, it is true that my holistic organization gives me the causal power to make such decisions as choosing tea over coffee. This agency is synonymous with an ability to navigate existence toward goal states that are not currently the case. If I were constantly being brought cups of tea, then I wouldn't need to set such a goal, but given that this is unfortunately not the case, in many situations I must set the goal and walk to the kitchen and make it myself to make that fantasy a reality.

This is the capacity for counterfactuality I touched upon earlier. Agents have the ability to reject reality as it currently is and so navigate to goal states that are other than one's current situation. All living systems must do this continually to achieve the goal of staying orderly. I believe it is this capacity for navigation that makes living things different from nonconscious wholes. It is what makes them epistemic agents, replete with the epistemic space we call awareness. I mentioned earlier that the process of evolution can also be modeled using the FEP, but in addition to not functioning as an integrated whole that could support a single perspective, it does not show agency of this kind. Unlike evolution and pendulums, we carve out space for ourselves in reality, resisting it as it is and dreaming of other ways it could be. Consciousness is a space of potential that is produced by this friction with reality, a bubble of possibilities that arises from the rejection of the flow of existence exactly as it is.

Locally resisting the second law sets up the organism-environment interaction and establishes the epistemic space of awareness. While we are alive, the extent to which we resist reality then serves to produce the contents of consciousness. In most forms of meditation, we drop our resistance to what is and for a time surrender this struggle with existence. In states of deep, meditative absorption, the contents of consciousness get exceedingly minimal, as we would expect if our experiences are synonymous with our capacity to resist the way things are. The epistemic space of awareness does not diminish during such experiences, however, in keeping with it being the result of our simply being alive. Even here, however, awareness is based in a kind of resistance, the organismic drive to stay orderly rather than succumb to the second law.

Where does the cutoff between life and nonlife occur? Viruses are typically held up as the paradigmatic example of something that exists at the boundary of life and nonlife. Viruses consist of strands of genetic material that piggyback on the life process to perpetuate themselves. They lack the capacity for agency that I examine here, instead appearing to simply respond mechanistically to their immediate circumstances.

Might we be able to create borderline cases of lifelike things that have a holistic organization yet do not navigate the world in order to maintain themselves? I suspect for this to be the case, we would have to continually pamper this molecular creation to keep it together, as it would not be able to perform this task itself. I believe there is a reason that we do not see borderline cases like this in the wild. While there is no hard boundary between chemistry and biology, we can imagine that there is a discontinuity of a kind. In dynamical systems theory, an attractor is a stable pattern of behavior that systems can evolve toward. This tendency to evolve toward a specific attractor and to settle into it is often visualized as a landscape, with attractor basins being depicted as dips in the landscape that the system can fall into. Life appears to exist in a basin of this kind.

While the landscape between life and nonlife is continuous, it is not linear. I suggest that the way of being that we call life is not merely an example of what happens when chemistry becomes gradually more and more complex. It represents a mode of organization qualitatively different from other kinds of complex chemistry. It is this disjunction that makes it possible for life to be associated with emergent phenomena like consciousness. Any borderline phenomenon we create will likely settle very rapidly into either the life or nonlife attractor basins on this landscape, determining whether we can say they are conscious or not. This is the same dynamic we observe when living things go from alive to dead, a transition that we assume also results in a sudden shift from consciousness being associated with that particular body to it no longer being present.

I cannot say for certain that we won't decide that there are phenomena more simple than life that we should also consider as being conscious, but the important thing is that this is the direction of travel that the debate should be focused on when considering consciousness, the direction beyond the living, not in the current direction of narrowing the source of consciousness down within the family tree of life to creatures sufficiently like us.

WHAT IS IT LIKE TO BE A BAT?

Theories of consciousness inevitably carry implications regarding which things are conscious, raising the issue of their ethical status, given that things that are conscious are things that may be capable of suffering. IIT holds that technology like the iPhone or the internet are sentient, that it feels like something to be a storm of Tweets or a breaking news story as new content cascades across the web. Panpsychism holds that even subatomic particles have experiences, leading one to wonder about the ethics of particle accelerators that smash these little sentient beings into each other.

Such odd implications are no reason to dismiss a theory of consciousness. As our understanding of our place in nature has progressed, we have been struck by a mind-boggling array of strange revelations. Our understanding of the microbiome has revealed that the majority of cells in your body are not human but bacteria, which are from an entirely different kingdom of life from humans. Without them, however, humans could not exist. Darwin's work revealed that we are related to all other living things and that each of us shares a direct ancestor with a banana. Research in cosmology shows us to be an unfathomably small part of the vast grandeur of the universe. Science illuminates very strange aspects of our existential situation.

According to the view I present, every living system has some form of experience. It is important to say that a worm would not be cogitating as you are now. It would not be capable of engaging in deep thought of any kind. To say that a worm is conscious is simply to hold that there is a distinct experience of being the worm. The touch of the soil on its skin feels a certain way, much as it does to us. This would hold true for single-celled organisms like bacteria that also navigate the world through their senses.

As I write this, a group of scientists have just released "The New York Declaration on Animal Consciousness,"[19] which advocates for a shift away from human exceptionalism and toward zoopsychism, the idea that all animals possess consciousness. Signed by leading figures in the field whose work has been explored here, including, among others, David Chalmers, Christof Koch, and Anil Seth, the declaration states that "there is strong scientific support for attributions of conscious experience to other mammals and to birds" and that "the empirical evidence indicates at least a realistic possibility of conscious experience in all vertebrates (including reptiles, amphibians, and fishes) and many invertebrates (including, at minimum, cephalopod mollusks, decapod crustaceans, and insects)." It is worth noting that this list still does not include any life-forms without nervous systems. Over the last several hundred years we have seen the Western science and philosophy shift from the assumption that only humans are conscious to an ever-expanding recognition of the number of living things that are capable of feeling. Perhaps we will see a similar declaration on the evidence for biopsychism in years to come.

According to this biopsychic perspective, plants, too, would have their own unknowable version of feeling. Rest assured, however, the apple tree isn't in pain when you pick its fruit. Aversive sensations like pain are synonymous with the organism reacting in a way that attempts to avoid that stimulus. If you raise my temperature to a lethal level or mortally dehydrate me, I will consciously suffer and will do everything in my power to change the situation to one that is more favorable to my survival. Inject me with a poison that puts me to sleep and stops my heart, and I will not suffer—the structure of my embodied responses didn't evolve to be able to prevent such a scenario. Fruit evolved to be eaten, and as a result, the fruit tree would have no reason to react negatively to its fruit being picked. Dry out the soil around the fruit tree's roots, however, and it may experience a negative state that may feel something like thirst as it reaches its roots down toward the water table.

Plants and bacteria need not share the same structure of consciousness as us. It's simply not necessary for their survival. We experience a self at the center of our experience that has its own personal narrative, but there is no reason to believe anything like this to be possible for such organisms. A sensation like pain, however, is so fundamentally tied up with this process of persisting by avoiding damage that it may be shared by all living things.

DOES AN AMOEBA HAVE BUDDHA NATURE?

Earlier I explained that conscious content depends on how thoroughly we resist reality in this moment. In states of deep, meditative absorption, conscious content is minimal, leaving the light of awareness to shine forth and to be recognized. The idea that conscious contents can vary in this way opens the door to the possibility that unicellular organisms do not experience very much at all, if we think of them as existing in highly predictable circumstances and surviving through very stereotyped behavioral routines.

If this proposal is correct, then it may be the case that simpler organisms exist in a kind of flow state much of the time, with little conscious content coming online. Rich content may have only developed with multicellularity, only to go into overdrive with nervous systems. The deep flexibility that nervous systems endow us with turns awareness of thriving versus suffering into a symphony of experience, of conjured worlds that we imagine ourselves to inhabit. There is no reason to believe an amoeba would be capable of anything of this kind, yet I suspect that the light of awareness has been ever present since the first organism came into existence, kindling a light in the darkness of reality. Fast-forward through evolution, and whether your experience is as a plankton, a fly, a redwood, an eagle, a wolf, or a human, all these experiential worlds play out in the same formless awareness. Still, there is a chance that the content of consciousness is minimal in simpler organisms. While this may reassure those who want to cling on to human exceptionalism, there is a big catch: We should think of such beings as effectively being enlightened.

Mahayana Buddhism teaches that the enlightened nature of the Buddha's mind is always present in the mind of every sentient creature, but it becomes obscured so that it is not recognized. In the terminology I use here, this claim can be interpreted as referring to the distinction I make between the conscious contents and the space of formless awareness. Formless awareness is the naturally self-luminous awake Buddha mind; it is Buddha nature.

With the space of awareness being established through the interaction with the environment in the service of survival, there is every reason to believe it is necessarily fully present in all living systems. The systems that do not struggle with existence as much as we do may have brightly shining minds without much going on inside. I suspect, however, that if we could put ourselves in their tiny shoes, rather than being enlightened or entirely unconscious, we might find that they live rich and ambitious lives with plenty of conscious content, but this is certainly debatable.

MULTIPLE MINDS

I am not one and simple, but complex and many.—VIRGINIA WOOLF[20]

I contain multitudes.—WALT WHITMAN[21]

While we typically think of ourselves as being a single thing, this is not quite accurate. As an organism, you are a symbiote, a collaboration between creatures from different kingdoms of the living world. If single-celled organisms are conscious, then does this mean the cells of the body are conscious, too? For the autonomous bacterial organisms that form a significant part of your body, there is every reason to believe they are associated with experience, according to the perspective offered here.

What about the human cells of the body: for example, a bone cell? Due to the manner in which we develop from a single fertilized egg, an integrative continuity unfolds in our development that results in these cells not functioning by the same principles as single-celled organisms. Given that many survival functions are centralized, bone cells, for example, may be able to function in such a predictable manner that there is minimal to no conscious content associated with the processes by which they maintain their individual order. If it is the case that some cells come to surrender much of their order-maintaining dynamics and are maintained by processes that occur as a part of the greater whole, then their conscious content may be minimal.

Bone cells do appear to have some cognitive capacities and have been found to exhibit foundational forms of memory, such as habituation and sensitization. These are nonassociative forms of memory, however. It is through making associations that we can decouple ourselves from the way the world currently is and become agents that can execute actions that move us toward counterfactual possibilities. Where bodily cells have surrendered this agency to be part of the greater whole, we may expect them to be like the enlightened amoeba, with formless awareness fully present but minimal content.

The other possibility is that they do have experiential worlds all their own, ones that are simply inaccessible from your perspective in the brain. Just as your inability to access my consciousness in this moment doesn't preclude me from being conscious, the same may be true for the subsystems of your organism. As odd as that sounds, there is nothing that makes it an impossibility. Only a deeper working out of the principles of predictive processing that underlie conscious content in the human brain and the subsequent exploration of whether such dynamics exist in these systems will be able to determine this one way or the other. Beyond this, to thoroughly answer the question of where different islands of consciousness exist in any given organism will require further progress in the science of the multiscale dynamical organization of living systems. Such study has been neglected in favor of the reductionist program of genetics, but a shift to a focus on the study of wholes holds promise for understanding development, the regeneration of body parts, and cancer.

Your brain is composed of multiple subsystems scaffolded on top of each other. Could it be that each of these systems has their own perspective on the world? Consider the neurobiology of pain. When you burn your hand, two pain pathways are recruited. One causes quick withdrawal without you being consciously aware of it, while the other, slower pathway leads to conscious pain emerging once the hand has been withdrawn. How can we account for the seeming lack of conscious experience being associated with the rapid hand-withdrawal reflex? One possibility is that the lack of counterfactual depth associated with the hard-wired reflex response leads to it not contributing to the conscious content. That is, consciousness is not necessary for the immediate retraction of the limb but is brought online for the longer-term attending to the damaged hand. Another possibility is that this circuit is associated with its own experience that is simply inaccessible for the system you identify with in the cortex, much as my experience is inaccessible to you. I call the system of consciousness that you identify with, the one in which thoughts and a personal narrative can arise, the conscious-self system. The role of the cortex in the flexible guiding of behavior may be a functional one that, rather than underwriting which contents come into existence, merely controls which information streams are recruited for behavior and become conscious from its perspective.

Earlier I explained that certain individuals with injuries to their visual cortex feel that they lose the conscious experience of vision but can still perform visually guided behaviors, such as identifying the location of an object in the blind region of their visual field without "seeing" it (blindsight, discussed in chapter 8).[22] The neurobiological explanation for this is that a secondary, more ancient pathway via the superior colliculus guides their behavior. In line with the two options provided here, we could argue that the relatively reflexive guidance of visual behavior performed by the superior colliculus leads to it not contributing to conscious content or that it is associated with a visual experience that is inaccessible from the perspective of the conscious-self system in the cortex. Crocodiles possess a comparable structure to the superior colliculus called the optic tectum, and the way we choose to answer this question informs how we think about the capacity of such creatures to experience. Is the crocodile an unfeeling killing machine operating on a kind of unconscious blindsight, with eyes that do not actually see, or does it experience the world around it? I'm inclined toward the latter. If this is the case, then the conscious-self system—the cortical system we identify with—is like a monkey riding on the crocodilian brain that is our superior colliculus, to offer a simplified analogy. We might just be a many-minded menagerie in one physical being, with different neural structures possessing their own mutually inaccessible perspectives on the world.

COULD ARTIFICIAL INTELLIGENCE BECOME CONSCIOUS?

In 2022, a software engineer working for Google, Blake Lemoine, announced to the world that he believed their AI chatbot LaMDA had become conscious. I reached out to Blake after he appeared on the news for several short-form interviews and became the first person to sit down with him for a long-form public conversation on the issue, on my consciousness-themed podcast *Living Mirrors*. While dismissed by many, Blake's stance was aligned with a mainstream position in consciousness science that I explored earlier: computational functionalism. According to this perspective, consciousness is the product of computations performed by the brain and could therefore also be created by a sufficiently complex computing machine. Blake felt that the team at Google had become the first people in history to cross this threshold and develop such a conscious computational machine, and he felt it his ethical duty to alert the world to this possibility.

So what about machines? Could they ever become conscious? This thought fascinates many and not just those who fear that films like *The Terminator* or *The Matrix* may one day become reality. It is also of huge ethical importance, as creating machines that can feel may bring vast amounts of suffering into existence. In addition to this concern and the thrilling sensation that we may one day be like gods ourselves, capable of creating feeling things, it proves a uniquely powerful testing ground for ideas about the mind. It's all well and good to speculate how physical stuff might be related to the mind, but if one's ideas are good enough, then they should be able to cross over from theory into engineering. What are the precise design principles that would be necessary to create a conscious being?

The core idea of this book is that consciousness is a relational dynamic that is a core aspect of the emergent self-organizing, self-perpetuating, energy-harvesting systems we call living things. No computer can implement these dynamics, as this would be a model or simulation of life and consciousness but not life itself. Machines are abstractions made concrete, things designed by an intelligence like us and set in motion. As such, they have no need to simulate the world around them as life does. If we created a machine that had the core features of life, the same autonomy and evolvability, then we could say that it would be both alive and conscious. Current computing, however, is not capable of generating a conscious being. A simulated mind becoming conscious would be like a computer simulation of a rainstorm getting the inside of the computer wet.

According to the perspective I present here, machines as we currently think of them could never be conscious. A machine is an artifact designed by an outside intelligence. Life, however, has to pull itself up by its bootstraps in order to exist. It is this self-authoring that necessitates consciousness in all living things. They must simulate the world beyond their boundaries in order to anticipate events around them and thereby survive. Living things do not have the luxury of a programmer to install their behavioral routines; they must invent them wholesale from within. Living things are not machines; we make machines as extensions of the problem-solving behavior that we engage in as biological agents. There is no way to think of machines apart from life. Our first-person position on the world, as systems that can care, understand, and act on the world as a result of our survival dynamics, is inescapable.

NATURALIZING SOUL AND SPIRIT

And if the body were not the soul, what is the soul?—WALT WHITMAN[23]

This equating of a relational mode of organization found in living things with consciousness that is offered here finds a resonance in the work of Aristotle. In his work *De Anima* (*On the Soul*), Aristotle suggests that the relationship between body and soul is like that between the wax of a candle and its shape: The wax is analogous to matter, while the superimposed form reflects the soul.[24] Soul here is understood as both that which makes us alive and that which endows us with the capacity to sense, think, and reflect, pointing to a connection between life and experience. For Aristotle, the soul is to be understood as a mode of organization. We can equate this with the emergent mode of systemic organization that channels matter and energy into a coherent whole, capable of both living and feeling.

Given this link between life and mind, however, this mode of organization cannot be interpreted as an immortal soul but is in fact a very mortal one. While our individual minds may be transient, the space of formless awareness offers a window through which we can perceive our continuity with the rest of reality while we are alive. In recognizing this continuity, we can come to identify more with the fact of being than with the passing form that we consider to be ourselves. In doing so, we end up feeling that our true natures will survive our individual deaths, as being itself is common to all that exists. This is a naturalistic conception of immortality, one that arises out of using one's individual awareness to recognize one's deepest nature as being the same thing as existence itself.

I suspect it is this experience that originally led to the idea of each of us possessing an immortal soul. If formless awareness is brought into existence through organism-environment interactions, then we can see why the recognition of formless awareness often leads to the feeling of being part of a unitive reality. Imagine if your eye were to believe it was responsible for seeing all by itself. If the eye were to suddenly realize that it is only one point in a larger process that involves photons on one side and the brain on the other, then it might realize that it is really the whole organism that sees, not the eye. Each of our consciousnesses is like the eye in this example, with the organism corresponding to the entire universe. Existence is a seamless whole that opens its eyes and sees through us, and conversely, formless awareness acts as an aperture through which we can come to know ourselves as the whole of existence. We can see here why Descartes believed the human mind to be based in a divine soul. The core of consciousness does indeed relate to "God," as the word *divine* suggests, if we understand "God" to be the nonconceptual totality of existence.

The process-relational neutral monism offered here also provides a way of making sense of the words *spirit* and *spirituality*. *Spirit* comes from the Latin word *spiritus* meaning *breath*, again pointing to a connection between consciousness and the living, breathing body. This connection is found in multiple languages, with the Greek words *psyche* and *pneuma*, the Sanskrit word *atman*, the Latin word *anima*, and the Hebrew word *ruach*, all meaning both *breath* and *spirit* or *soul*. The idea of breath nicely captures our process nature: that is, that we are a *process* occurring in the natural world rather than a mysteriously animated separate piece of matter. In mystical states we see through the illusion of solid substance that keeps us from appreciating this process nature in ourselves and in wider reality. What's more, we see through the illusion of being a separate self that is shut off from the rest of existence. The idea of spirituality then can be understood as the practice of orienting to this process connection between oneself and the rest of existence, of seeing that the breath of life is not something separate from the currents of the rest of existence.

This may sound like a romantic vision of our place in existence, but it may in fact be quite confronting to the parts of ourselves that want to find a deeper meaning and purpose in reality. While the universe can be understood as waking up through living systems like us, there is no reason to believe that there is any deeper meaning behind this. It is not that the universe wants to wake up or that it has been trying to bring this about. Existence simply unfolds in a lawful manner as described by science. It is not necessarily happening for any ultimate reason, and as far as we can tell, it is not heading to any inherently meaningful destination. Meaning and reasons arise with minds like ours, but they are thankfully absent from reality at the more fundamental level. This perspective invites us to settle into the freedom that is experienced when we feel what it is to exist beyond meaning and purpose.

WHAT'S GOING ON?

The goal of life is to make your heartbeat match the beat of the universe, to match your nature with Nature.—JOSEPH CAMPBELL[25]

I've explored a lot of nuance and detail from across multiple fields of inquiry to come to what I believe is quite a simple picture of what is going on. All that exists is a single relational process that we call reality. Within this process, the dynamics of selection pull complex forms into existence across multiple scales, with reality hanging together as a single whole. It may feel like you are a truly autonomous agent with free will, but your decision making is being effortlessly sculpted by the selection algorithm in much the same way that evolution bequeathed you with a human form without you having to lift a finger. In such a picture of reality, there is not much point in attempting to struggle unnecessarily with the wider world. Instead, the unfolding of existence invites us in every moment to relax into it, to sit back and simply enjoy the eternal ride of forms being pulled into existence out of the formless space of potential.

This may be what is going on in all of reality, but what is going on for you in this moment? Experiential content is continually arising and passing away in the epistemic space of formless awareness that is a feature of the organism-environment complex where you are, the complex we call your body and your immediate environment. An image of a self is constructed in this awareness like a standing wave in this relentless flow of experience, but it is only an impression. There is no "thing" there that you could be, separate from the ever-changing contents of consciousness.

If we shift our focus to the epistemic space of awareness, then perhaps we'll be on firmer ground. Is this awareness what you are? This idea is just that: an idea, a thought. Prior to that thought, the awareness just is. There is truly no need to add the imagined layer of identity that insists that you must be something other than me and everything else in existence. We can talk about two islands of awareness existing where our two bodies are, but they are not different types of awareness; they are the same thing. Despite many examples of flames existing, there is only one phenomenon of fire, and the same is true for awareness. There is only one phenomenon called awareness. This unity does not mean that it is a continuous field, in the same way that different flames do not have to join together to all be considered examples of fire. What it does mean, however, is that the most fundamental aspect of your mind lies beyond your personal identity and is as free of concerns of me versus you as the flame is.

If you are anything—and that's a big *if*—then you are the same thing as what is. While awareness is part of what is, it is not the whole thing. Awareness exists as a particular part of reality where living things interact with their environment. Awareness does not exist apart from the greater process of existence, however. It arises as a phenomenon within and through the activity of the greater whole. This greater whole is the relational process explored in chapters 4 and 5, an interdependent web of forms perpetually arising out of and returning to a space of formlessness. The story of physics deals with the network of empty occurrences we call matter, a process that forms the earth and your body. A similar interdependent process in belief space weaves together your coherent experience of existence in consciousness. These phenomena meet in the process that is the living body. Matter provides the building blocks, and a feeling mind emerges as a holistic mode of organization, through which information is channeled to make survival possible. We are not a ghost in a machine. Rather, it is as if a vortex of machine parts had come together to form an emergent ghost.

The worldview presented here offers a way to synthesize different perspectives on consciousness. The story I have told should sit well with physicalists, as consciousness is held to be nothing but a certain arrangement of physical processes. It is not something truly distinct from the operation of the physical world. It should satisfy panpsychists, as the essence of mind—relationality—does indeed exist in matter, which could be said to possess a kind of protoconsciousness in the sense that it provides a foundation out of which consciousness can emerge but not one that by itself possesses the capacity to experience. Idealists should also be happy with this view, as it is one where there is no solid material substance in existence, and the essence of consciousness—both its formless ground and the relational forms that we experience as its content—is what all reality is made of.

Ironically, given that I am a neuroscientist, the main view that is not compatible with this picture is the idea that brains or nervous systems are necessary for consciousness to come into existence. This view is what leads to the hard problem, and it creates philosophical inconsistencies that are evidence of its lack of viability as a perspective. The story presented in this book can account for why this neurocentric view is widely held, however, as the brain is held onto as the last holdout for human exceptionalism and the imagined superiority of our species over the rest of nature. With any luck, we will move beyond this supremacist perspective sooner rather than later. Failing to do so not only prevents us from understanding ourselves and our place in the world more deeply but also may be the very undoing of the modern world.

EPILOGUE

Healing the Divide between Nature and Consciousness

Everything the Power of the World does is done in a circle. The sky is round, and I have heard that the earth is round like a ball, and so are all the stars. The wind, in its greatest power, whirls. Birds make their nests in circles, for theirs is the same religion as ours. —BLACK ELK[1]

FROM SEPARATION TO A RELATIONAL WORLDVIEW

The core claim of this book is that all life is conscious. The mainstream assumption is that the capacity of other creatures to experience exists by virtue of their similarity to us; apes and perhaps dogs are permitted to feel, but insects and fungi are suspect. I showed how this stance of human exceptionalism has not been held throughout all of human history. Animistic visions of the world in which we are fully part of the relational web of nature appear to be more common to humans than the idea of our separation from and superiority to the natural world. We have here two ideas of the world and our place within it: One is a relational view; the other is a view of human supremacy and separation.

The view of human supremacy went into overdrive with the emergence of our current capitalistic economic system. This way of organizing our material relationships seeks to reduce the multiplicitous value of ourselves and our world into a single monetary quantity based on its instrumental value. It is a powerful framework for extracting and concentrating this specific form of value, but it does so at the expense of everything else we care about, from our well-being to the climate. We all value having a habitable biosphere, but this value is not incorporated into the way that we conduct our economic activity, as a healthy environment cannot be exploited to give anyone greater power in the short term.

The ideas we have about the world, the way we perceive it, and the way we interact with it all shape each other and in turn shape us. It is my contention that our beliefs in human supremacy, our perception of the living world as other, and our extractive and self-destructive economic practices come as a package deal of ignorance. This is not the only way of thinking about, perceiving, and relating to the wider living world, however. All three of these issues can be addressed by shifting away from a worldview of separation, where what is not "me" is considered fair game for violent exploitation and instrumentalization, and toward an interconnected relational worldview, where we interact with each other and nature through nonviolent, consensual, reciprocal ways of acting. We can begin to disabuse ourselves of our current delusional worldview by seeing that our very ability to take in existence in this moment through the gift of experience is possible because we are living things, like all other living things, rather than being due to our being alienated, disembodied computations running on brains, separate from an objectified natural world.

Our ideas about the world are unlikely to change through reason alone. We have seen how all mental activity exists in the service of survival. We form our beliefs not only out of a need to be in touch with reality but also out of a need to protect our ways of life that support our continued existence. According to the free-energy principle (FEP), emotionally motivated reasoning and confirmation bias are fundamental strategies of survival. Life cannot know what is objectively the best way for it to survive in its specific environment, so once it finds a strategy that works, it sticks to its guns as much as it can, as this way of being has kept it alive until now. This is a challenge we face as survival-oriented living things that are interested in understanding ourselves. If we were disembodied rational intellects, then we might be able to simply submit the available information to analysis and to accept whatever conclusion this process produced. This is not our situation, however. We very much have skin in the game when it comes to which ideas we are willing to accept as true.

DEEP-ECOLOGICAL HUMANISM

Surely, even you, at times, have felt the grand array; the swelling presence, and the chorus, crowding out your solo voice. —DAVID WHYTE[2]

The idea that living things are conscious has direct implications for the ethics of exploiting the natural world. We have a natural capacity for empathy, and we intuitively reject the domination of those we consider sufficiently like us to have moral worth. We see this in the way that slavery was justified in the minds of slaveowners through the dehumanization of the enslaved. Empathy is a psychological process that exists to mobilize behaviors directed at relieving the suffering of those we empathize with. Seeing the person one has enslaved as human like oneself would have produced a profound cognitive dissonance in the mind of the enslaver, a dissonance that needed to be resolved in one way or another. It is no coincidence that Descartes's picture of reality in which nature is unconscious arose alongside an economy based around the exploitation of nature and the enslavement of other humans. By turning these subjects into objects, empathy could be withdrawn, and those deemed less than human could be exploited for economic gain, a process that continues to this day. The mind-body problem is really the subject-object problem in a world that assumes many other living things to be exploitable objects.

A lifestyle based in the domination of nature is one that emphasizes short-term benefit to oneself over long-term flourishing, as we see with the damage being done to the climate. This kind of self-centered behavior is often presented as the norm, a simple result of Darwinian evolution, but this idea is ahistorical.[3] There is wide-ranging evidence for reciprocal economies of care in Indigenous communities that seek survival through mutual support rather than competition. Several decades after the publication of Darwin's *On the Origin of Species*, Russian naturalist and anarchist political thinker Peter Kropotkin wrote a work entitled *Mutual Aid: A Factor of Evolution*, which documents this principle of survival via cooperation rather than competition in nature, a phenomenon we see throughout the natural world, from such symbiotes as lichen to the mitochondria in your cells that teamed up with ancient protozoa to the benefit of both.[4]

What would lead a culture to withdraw its empathy in order to dominate and control other living things to the extreme extent that is the norm today? Dominating others to accumulate power for oneself can be thought of as the lowest-common-denominator survival strategy that we rely on in moments of extreme stress. Those who rescue drowning people are often told to be aware that the drowning person may instinctively push their would-be rescuer under water in their blind panic to stay afloat. Europe by the time of Descartes was not a place full of people pursuing emotional flourishing. After millennia of war and conquest, as well as such epidemics as the Black Death, it is not surprising that the culture carried enough collective trauma that the landowners acted like the drowning individual, enslaving the natural world, the majority of the local population, and other continents full of people to their material benefit to keep the fear and tangible reality of not surviving themselves at bay. This is not to justify or excuse slavery, imperialism, and the destruction of the natural world but is an attempt to understand in part why they came into existence and continue to exist today.

COLLECTIVE TRAUMA

We do not see things as they are, we see them as we are.—ANAÏS NIN[5]

Science and technology can be understood as an extension of the Darwinian survival strategy of life. Technological society extends its control over the natural world by creating, testing, and retaining theories that match up with reality, the same algorithm that underlies the survival and evolution of life. We could argue that the spread of technological civilization over the globe was an inevitability due to the brutal logic of selection, where the societies best at surviving survive and spread. The issue is that this survival strategy can only work in a limited time frame, as it relies on the autocannibalism of the biosphere that we all depend on for our existence. In the long term, evolution favors survival of the stable, which may be why, prior to European colonization, Indigenous societies of peoples who lived in a sustainable relationship with nature were present over such a large extent of the globe. Consciously organizing around the wisdom of long-term sustainability does not put us in opposition to these powerful societal selection dynamics; it places us to catch the wave of survival while more fragile ways of life based around short-term domination sink beneath the waves. Such ways of life are called sustainable for a reason.

If collapse of this kind is the inevitable destination of our current way of life, then why would a culture adopt such an unsustainable and unbalanced mode of existence in the first place? When death looms large, we understandably become focused on our own survival. The imminent possibility of death need not be in the cards for this process to occur, however. The Bayesian dynamics that underpin experience consist of updating one's existing beliefs rather than seeing reality as it is. As a result, past challenges to our survival, well-being, or agency can cloud how we see the world. We call these traumas. We now know that such traumas need not be experienced in one's personal lifetime to affect one's behavior and can actually be inherited from previous generations.[6] What's more, they can play out at a collective level, influencing culture and societies at large.[7] Collective trauma is arguably at the root of our global societal pathology.

Is it necessary to keep living in this way at this point in human history? We live in an uncertain world where self-protection is always necessary, yet without examining these traumas, we cannot assess which self-protective strategies are adaptive and which are outdated. Perhaps in medieval Europe a conquer-or-be-conquered strategy of preemptive domination was adaptive from the perspective of the conquerors, but in our modern interconnected world, it may be more neurotic and dysfunctional rather than adaptive. We see this in the many social and environmental ills wreaked upon the world.

Thankfully, we do have the means to update our beliefs and let go of those that we deem to be outdated. This requires us to turn toward the contents of our embodied conscious experience in the present moment to confront the beliefs through which we interpret the world. We typically function by avoiding such confrontation, instead letting our attention pull us into thoughts of the future or the past rather than feeling what is here now. Becoming aware of what is present now is the capacity we call mindfulness, and it is the fundamental skill that is necessary for the shift in orientation described here.

Sometimes this act of becoming mindful is enough to update a belief. In such cases the act of simply stopping results in a kind of homeostatic rebalancing, where the belief can be revised in light of evidence. Imagine a child who becomes very fearful while watching a scary movie because they become so engrossed in what is portrayed on the screen. By simply reassuring them that it is a film and directing their attention to this fact, their belief that they need to mobilize a fear response can be immediately updated, and they may calm down. Similarly, for an adult who experiences mild social anxiety, mindfully orienting to their physical safety in the present moment may sufficiently update the belief that they are under threat in that moment in order to relieve their anxiety.

Beliefs formed during childhood or as the result of a traumatic event feel particularly significant for our survival and so are typically not updated this easily. In such cases and for certain individuals, psychedelic medicine can be used to put the brain in a state where belief updating is made possible.[8] The result of this internal psychological work is that it becomes possible to shift from a dynamic of avoidance, where we feel disconnected from aspects of our own experience and from the world, toward connection with ourselves and the rest of existence. This shift creates a buffer of awareness between events unfolding in the world and our reactions to them. With greater mindfulness and less unconscious reactivity, we are better placed to make decisions that will contribute to greater collective flourishing in the long term.

Wisdom is the term we use for the ability to focus on the bigger picture. Our culture does not emphasize wisdom, instead choosing to focus on short-term material gain. Cultures that do emphasize wisdom and focus on the long term typically practice a reciprocal engagement with the natural world. For example, the Haudenosaunee or Iroquois peoples of North America both advocate for factoring in the effect of any decision on descendants seven generations into the future and practice principles of reciprocity in their relationship with nature, although in effect these practices are not separate but support each other.

As we concern ourselves with the bigger picture of our situation, we see the feedback loops of the biosphere as relevant to ourselves. With this expanded vision comes an expansion in our sense of self. This feeling of connection with the biosphere can be contrasted with the current dominant view of separation. In identifying with other living things, we expand our circle of moral concern to include anything that can suffer. This is aligned with the perspective of deep ecology, which sees humans as just one of many species on our planet that are of equal value. If we take seriously the idea that all living things have the capacity for experience, then we are faced with trade-offs in the suffering we cause. For example, we may be inflicting pain on the thousands of bacterial organisms we kill whenever we wash our hands. This is an uncomfortable thought, and a common instinct is to reject it on an emotional level and insist we must come up with an explanation of why this could not possibly be the case. If we entertain the possibility that it might be true, then what would it mean for how we live and for leading an ethical life?

This picture invites us into a reckoning with the inevitability of suffering and death for all living things. To favor the lives of the bacteria over oneself would lead to an unsustainable world for humans, in which huge numbers of us would die of preventable disease. We are currently biased far in the other direction, in prioritizing human dominance and not necessarily human flourishing at all costs. I believe the solution is a balance between the two, a kind of deep-ecological humanism that would accept our innate preference for permitting human flourishing over other organisms but would couch it in the wisdom of trying to minimize suffering for all sentient life. If our culture came to a consensus on engaging with the world in this way, most likely nothing would change in our practices when it comes to bacteria, but we would surely see a decrease in unnecessarily cruel practices, such as factory farming.

This interplay of human concerns and a wider concern for the more-than-human world is reminiscent of animism. While anthropologists in the 1800s initially presented animism as a belief in spirits pervading the living and nonliving world, modern scholars have powerfully argued that animism should be more accurately understood as ways of relating respectfully to the rest of the natural world. An animist worldview does not require humans to disregard their own interests but acknowledges how human actions affect other beings with their own concerns. Making a practice of taking these other perspectives into consideration is at the core of animists' sustainable relationship with the world.

This respect-based way of relating stands in stark contrast to the dynamics of domination and disconnection that characterize the modern world. Modernity is fundamentally built on the disrespect of the natural world and often of our fellow humans. Human supremacy (including human exceptionalism in consciousness science) and interhuman ideologies of supremacy pervade modern societies. Ultimately, divesting from supremacist psychology in all its many guises lies at the core of rediscovering our ecological nature and embeddedness in the wider world. A shift from disrespectful domination and exploitation and toward respectful and reciprocal relationality would serve to address the root cause of ecological breakdown and the climate crisis—a life way born of our felt disconnection from the natural world.

FACING DEATH

Each wave is born and is going to die, but the water is free from birth and death.—THÍCH NHẤT HẠNH[9]

Our existential situation is one of precarity: The second law will always win in the end. Death is an ever-present possibility and something that many spend their lives turning away from. Flourishing requires us to face reality so that we can cocreate a world aligned with our highest values. Rather than turning toward uncomfortable truths about our situation, however, our current way of life is based around avoidance. Like someone soothing emotional pain through their addiction to a substance, we are addicted to fleeing connection with this experiential moment by distracting ourselves in thought, in hopes of things being different in the future. Our culture typically avoids facing the reality of our situation in this way, preferring to focus on the possibility of defying aging and death with technology or devastating the biosphere to satisfy our short-term desires. If this whole dynamic is so pathological, then why do we do it? If it's so much better to come into connection with oneself and the wider world, then why don't we simply do that? The proximate reason is that orienting to whatever it is in our experience we are seeking to avoid is always uncomfortable. Learning to become comfortable with discomfort is an essential component of orienting away from separation and toward a relational worldview. The ultimate cause of our avoidance, however, is the reality of death.

All the meaning that arises in our minds is scaffolded on the reality of and resistance to our inevitable deaths. The fact of death is what gives rise to the sense of being a self, separate from the rest of existence, a small, vulnerable being that must navigate in a hostile world that is other than itself. The fear of death plays an outsized role in modern society, as it is not actively reckoned with and resolved. As a result, our culture pushes us further and further into self-based psychology rather than the psychology of universal compassion. In *The Denial of Death*, Ernest Becker argues that much of culture arises as an effort to not face the reality of death.[10] I believe this to be correct. To see our relationship to the biosphere clearly and to behave toward it in a way that is more sane, we must confront the reality of death.

The good news is that, experientially, the dynamic of impermanence and loss that points to our inevitable deaths is an ever-present reality that can always be engaged with. This may not seem like good news to the part of you that is terrified by your own mortality, but the suggestion here is that there is a wisdom to engaging this experiential terrain that outweighs this fear. By embracing the experience of loss as a reality of life, we can bring ourselves into greater harmony with existence rather than acting out of a neurotic rejection of this aspect of reality as we are currently doing en masse.

We started out with a metaphysical picture of reality as being grounded in a formless nothingness. As a result, all forms are impermanent and return to this nothingness. By facing this fact of impermanence and becoming comfortable with the feelings that arise as a result, we find ourselves aligning further and further with the reality of this moment. What we find there appears to be an eternal, cyclical process. On the one side, there is the arising of the ever-blossoming forms of reality, and on the other is the inevitable decay and death of these forms. In not rejecting the side of death and decay, we find our hearts beating to the same rhythm as reality itself, open to the eternal flow between being and nonbeing. By seeing that the relative forms that arise in existence are always in a state of being born into this moment and simultaneously dying out of it, a deathless, unchanging absolute ground of formlessness is revealed that can support us in this confrontation with impermanence.

What I suggest here is not to become morbidly introspective about death but instead to take an unbiased interest in the nature of experience. When we do this, we face our experience of reality more and more, and we find great experiential freedom in this alignment with what is. On the journey we integrate different parts of our experience into a harmonious whole and taste ever-greater liberation from the painful struggle with existence that is our evolved default. What's more, this harmony inevitably radiates out into our relations with the wider world. In the beginning this project can feel entirely positive for some. For others, the need to face one's inner demons emerges quickly. In the end, however, for everyone who sticks with this journey, it will ultimately involve truly facing death and impermanence head-on. The choice is between the struggle of hiding this aspect of existence from yourself until it eventually catches up with you, perhaps during a midlife crisis or after an unexpected medical diagnosis, or choosing to face it.

This is not about finding a fixed solution or ending up at a perfect end state, but rather it is about consciously orienting in the direction of wisdom rather than from reflexive, fear-based instincts. It is a movement that is always available in our minds in every moment. All we need to do is to let go of distraction and open to the reality of our conscious experience as it is. Facing reality in this way takes the deepest courage imaginable. The journey is only possible with support and care, both from oneself and from others. It is arduous, but the pursuit of experiential truth is the deepest pursuit any of us can spend our lives undertaking. Going on this pilgrimage into the heart of the interplay of form and formlessness is truly the journey of a lifetime and may just be what is needed to rectify our current collective ignorance. If we can be courageous enough to make a home for ourselves here, at the very heart of existence, then we may find relief from our collective desire to objectify and oppress each other and the rest of the natural world. In doing so, we may come to see our situation and ourselves more clearly and to suffer less as a result.

ACKNOWLEDGMENTS

I am very grateful to live in a time when it is possible to stand on the shoulders of so many intellectual giants on the topic of consciousness. After having the initial insight into the connection between experience and life that led to this book, I was deeply reassured to find that others had intuited something akin to what I had. These inspirational theorists include Francisco Varela, Humberto Maturana, Evan Thompson, Lynn Margulis, Arthur Reber, Terrence Deacon, Mark Bickhard, Antonio Damasio, Alicia Juarrero, Alva Noë, Maxine Sheets-Johnstone, Anil Seth, Andy Clark, and Karl Friston, and I am sure there are others who are worthy of mention here. Thank you all for your contributions. I am also thankful for the scientists of previous generations who have sought to normalize the project of connecting the scientific with the spiritual, such as Fritjof Capra.

I am grateful to all the friends and colleagues with whom I discussed these ideas and who gave me feedback on the manuscript, in particular Shamil Chandaria, who is a true kindred spirit when it comes to these topics and has contributed significantly to many of the ideas I explore here. My appreciation goes out to Gioia De Franceschi, Fabian Peters, Louise Guidi, and Cathy Price for feedback on early drafts of this book and to Matt Segal and Graham Harvey for sharing their expert advice on specific topics covered in the book (panexperientialism and animism, respectively).

Many thanks to my agent, Don Fehr, for securing my first book deal, something I have dreamed of since I was a teenager. Many thanks to my editor, Jonathan Kurtz, for his enthusiasm for this project and for making this book a reality.

I am appreciative of those who influenced me on this path decades ago, my teachers at college, David Horner and Ian Yates, who facilitated the beginnings of my career in academia, and my mother, Dr. Jacqueline Elliot, who first pointed me in the direction of studying psychology. I would also like to acknowledge my grandmother, Josephine Kennard, who has been a guiding light throughout my life and a perpetual cheerleader for my endeavors.

Above all I am grateful to my incredible wife and inspiration, Rebecca, who, in addition to offering unfathomable emotional support in life, has been absolutely crucial in helping to shape these ideas. Her profound insight, razor-sharp intellect, creative genius, and spiritual depth all contributed to the thorough fleshing out of these ideas before they were committed to print. If you find value in this work, you would have found only a fraction of that value if it were not for her.

Finally, thanks to you for taking the time to explore these ideas. I find it profoundly moving that someone would dedicate hours of their life to exploring my perspective on our nature and the nature of our existential circumstance. I truly hope you find value in this work.

GLOSSARY

active inference: a theoretical approach to predictive processing that incorporates action

affordances: opportunities for action provided by the environment, perceived in relation to an organism's capabilities

animism: the belief that natural phenomena, such as animals, plants, and inanimate objects, are all capable of being seen as persons who should be related to with an attitude of respect; sometimes presented as the belief that all such phenomena possess an animating or spiritual essence

awareness: the experiential space in which conscious contents arise

Bayesian belief: a hypothesis about the state of the world, represented probabilistically

Bayes' theorem: a mathematical formula describing how to update the probability of a hypothesis being true in light of new evidence

biopsychism: the philosophical viewpoint that all living things are conscious

blindsight: a condition in which individuals with damage to certain visual areas of the brain are able to respond to visual stimuli without consciously perceiving them

brainstem: a region of the brain that regulates basic physiological functions, such as heartbeat, breathing, and physiological arousal

catalytic closure: the property of a chemical feedback loop whereby the set of reactions that make it up can catalyze themselves

classical conditioning: a form of learning in which an organism comes to associate a neutral stimulus with a biologically significant stimulus, resulting in a change in response to the neutral stimulus

cognition: the process of relating to a domain of significance through processes such as perception, learning, reasoning, and decision making

computation: the manipulation of symbols according to predefined rules so as to transform an input into a desired output

concepts: mental constructs used to structure our experience of the world

consciousness: the capacity to experience; the fact that it is "like something" to be the system or process that is conscious

constraints: limitations that shape the behavior of evolving processes

cortex: the outermost structure of the brain, responsible for higher cognitive functions

counterfactual: alternate possibilities that are different from the way something actually is

dualism: the philosophical idea that reality consists of two fundamentally different kinds of substance: matter and mind

eliminative materialism: the philosophical view that mental states, such as consciousness, do not exist and should be eliminated from scientific discourse

emptiness: a concept in Buddhist philosophy referring to the lack of independent existence in phenomena and the corresponding interdependence of all things

enactivism: a theoretical approach to cognition that emphasizes the embodied and situated nature of the mind

entropy: a measure of disorder or randomness in a system

epiphenomenal: referring to mental states or phenomena that are causally inert by-products of underlying physical processes

epistemic: relating to knowledge

folk psychology: commonsense understanding of the mind and behavior based in how they superficially appear to us (in contrast to understanding that relies on scientific or philosophical investigation)

free-energy principle (FEP): a theoretical framework in which living systems and the brain minimize a quantity called free energy to achieve their goals and survive over time

functionalism: the philosophical theory that mental states are defined by their functional roles, rather than their physical instantiation

hard problem of consciousness: a term coined by philosopher David Chalmers to refer to the challenge of explaining why and how consciousness arises from physical processes

homeostasis: the ability of living organisms to maintain their internal states within a range that is conducive to their continued survival

human exceptionalism: the belief in the privileged status of human beings, often regarding consciousness

idealism: the philosophical view that reality is fundamentally mental and that what we call matter is dependent on consciousness for its existence

illusionism: the philosophical position that consciousness is an illusion

information: the reduction in uncertainty produced by the receiving of a signal

integrated information: a measure of the extent to which a system's elements interact to produce an irreducible holistic organization

interoception: the ability to sense the internal state of one's body

metaphysics: the branch of philosophy that deals with the fundamental nature of reality

micropsychism: the view that consciousness is a property of fundamental physical entities, such as physical particles; a version of panpsychism

monism: the philosophical view that there is only one underlying substance or principle in reality

mysterianism: the philosophical position that consciousness may be inherently beyond human understanding

mystical experience: states of consciousness characterized by a sense of unity and interconnectedness

naturalism: the philosophical view that all phenomena can be explained in terms that are consistent with scientific investigation and without invoking supernatural explanations

neural correlates of consciousness: the neural processes associated with the contents of conscious experience

neutral monism: the philosophical view that reality consists of a single underlying substance or principle that is neither exclusively mental nor physical

nonconceptual: existing apart from the concepts that structure our experience

nonduality: the insight that reality is ultimately undivided and cannot be adequately described or understood in terms of separate or distinct entities

noumenal: relating to the realm of things as they are in themselves, beyond perception or experience (i.e., the phenomenal)

objective: relating to phenomena that are publicly observable and exist independently of one's experience of them; the opposite of subjective

objective idealism: the philosophical view that reality is ultimately mental but exists independently of our individual minds

ontology: the branch of philosophy concerned with the nature of being

panpsychism: the philosophical view that consciousness is a fundamental feature of the universe, present in all things to varying degrees

perception: the process of recognizing, interpreting, and organizing sensory information in order to apprehend the environment and often oneself within it

phenomenal: relating to felt experience, as opposed to noumenal reality

physicalism: the philosophical view that the aspect of reality studied by physicists is fundamental

prediction error: the discrepancy between the predicted outcome of a situation and the actual outcome, used in predictive processing to improve the accuracy of predictive models

predictive processing: a theoretical framework in which the brain generates predictions about incoming sensory information and updates these predictions based on prediction errors generated as a result of feedback from the environment

process: something that is fundamentally dynamic in nature; process explanations stand in opposition to substance explanations

protoconsciousness: the rudimentary forms of consciousness hypothesized in panpsychism to exist in simple physical systems

qualitative: relating to properties or attributes that are not quantifiable

quantitative: relating to the measurement or quantity of an object or phenomenon

reductionism: the philosophical approach that seeks to explain complex phenomena by reducing them to simpler, more fundamental components or processes

relational: relating to relations, connections, or interactions

representation: a symbol that carries encoded information about the world in a form that can be stored and manipulated

selection: the process by which certain traits or characteristics persist over time due to their advantageous effects on survival

solipsism: the philosophical belief that one's individual mind is the only thing that exists

subjective: relating to the private nature of experience, not directly observable by others; the opposite of objective

subjective idealism: the philosophical view that reality is ultimately mental and that the external world exists only as perceptions within individual minds

surprise: a measure of how unexpected a signal is in information theory; more surprising signals are less expected and therefore more informative, as they reduce more uncertainty

teleology: the philosophical study of purposes, goals, or ends and their role in explaining the properties of natural phenomena

transcendental idealism: a philosophical view that reality consists of both the phenomenal world of experience and the noumenal word in itself, as proposed by Immanuel Kant

universal Darwinism: the application of Darwinian principles of variation, selection, and retention to explain a wide range of phenomena beyond biological evolution

vitalism: the philosophical doctrine that living organisms are fundamentally different from nonliving matter and possess a unique "vital force" or principle that distinguishes them from inanimate processes

zoopsychism: the philosophical view that animals possess consciousness

NOTES

INTRODUCTION

1. T. Nagel, *What Is It Like to Be a Bat?* Language and Thought Series (Harvard University Press, 1980), 159–68.
2. C. McGinn, "Consciousness as Knowingness," *Monist* 91, no. 2 (2008): 237–49.
3. J. E. Cooke, "The Living Mirror Theory of Consciousness," *Journal of Consciousness Studies* 27, nos. 9–10 (2020): 127–47.
4. C. Sagan, *Cosmos* (Random House, 1980).

CHAPTER ONE

1. D. J. Chalmers, *The Conscious Mind: In Search of a Fundamental Theory* (Oxford Paperbacks, 1997).
2. G. Galilei, "The Assayer," in *Discoveries and Opinions of Galileo*, trans. S. Drake (Anchor, 1957).
3. F. Crick, *The Astonishing Hypothesis: The Scientific Search for the Soul* (Scribner, 1995).
4. Aristotle, *On the Parts of Animals*, trans. J. G. Lennox (Clarendon Press, 2002).
5. F. Crick and C. Koch, "Towards a Neurobiological Theory of Consciousness," *Seminars in Neuroscience* 2 (1990): 263–75.
6. D. A. Leopold and N. K. Logothetis, "Activity Changes in Early Visual Cortex Reflect Monkeys' Percepts during Binocular Rivalry," *Nature* 379, no. 6565 (1996): 549–53.
7. D. J. Chalmers, "Facing Up to the Problem of Consciousness," *Journal of Consciousness Studies* 2, no. 3 (1995): 200–219.
8. D. J. Chalmers, *The Conscious Mind: In Search of a Fundamental Theory* (Oxford Paperbacks, 1997).
9. P. M. Churchland, "Eliminative Materialism and the Propositional Attitudes," *Journal of Philosophy* 78, no. 2 (1981): 67–90.

10. D. C. Dennett, *Consciousness Explained* (Penguin UK, 1993); K. Frankish, "Illusionism as a Theory of Consciousness," *Journal of Consciousness Studies* 23, nos. 11–12 (2016): 11–39.

11. C. McGinn, "Can We Solve the Mind-Body Problem?" *Mind* 98, no. 391 (1989): 349–66.

12. P. De Laplace, *A Philosophical Essay on Probabilities* (Courier, 1995).

13. Quoted in A. Osborne, ed., *The Collected Works of Ramana Maharshi* (Weiser Books, 1997).

14. R. Descartes, *Meditations on First Philosophy*, trans. I. Johnston (Broadview Press, 2013).

15. G. Berkeley, *A Treatise Concerning the Principles of Human Knowledge* (1710; RS Bear, 1999).

16. B. Russell, "On the Nature of Acquaintance: II. Neutral Monism," *Monist* 24, no. 2 (1914): 161–87.

17. P. Goff, *Consciousness and Fundamental Reality* (Oxford University Press, 2017); P. Goff, *Galileo's Error: Foundations for a New Science of Consciousness* (Vintage, 2019); G. Strawson, *Realistic Monism: Why Physicalism Entails Panpsychism* (Oxford University Press, 2008).

18. G. Tononi, "An Information Integration Theory of Consciousness," *BMC Neuroscience* 5 (2004): 1–22; G. Tononi and C. Koch, "Consciousness: Here, There and Everywhere?" *Philosophical Transactions of the Royal Society B: Biological Sciences* 370, no. 1668 (2015): 20140167.

19. D. Hoffman, *The Case against Reality: Why Evolution Hid the Truth from Our Eyes* (W. W. Norton, 2019); D. D. Hoffman, M. Singh, and C. Prakash, "The Interface Theory of Perception," *Psychonomic Bulletin and Review* 22 (2015): 1480–1506.

20. C. McGinn, *The Problem of Consciousness* (John Wiley and Sons, 1993).

21. W. Sellars, "Philosophy and the Scientific Image of Man," *Frontiers of Science and Philosophy* 1 (1962): 35–78.

CHAPTER TWO

1. R. W. Emerson, *The Complete Works of Ralph Waldo Emerson*, vol. 5 (Wm. H. Wise, 1903).

2. A. Einstein, "Letter to Dr. Robert Marcus," Library of Consciousness, February 12, 1950, https://www.organism.earth/library/document/letter-to-dr-robert-marcus.

3. H. Päs, *The One: How an Ancient Idea Holds the Future of Physics* (Icon Books, 2023).

4. F. Capra, *The Tao of Physics: An Exploration of the Parallels between Modern Physics and Eastern Mysticism* (Shambhala, 2010).

5. A. M. Liberman, K. S. Harris, H. S. Hoffman, and B. C. Griffith, "The Discrimination of Speech Sounds within and across Phoneme Boundaries," *Journal of Experimental Psychology* 54, no. 5 (1957): 358.

6. H. Barlow, "Grandmother Cells, Symmetry, and Invariance: How the Term Arose and What the Facts Suggest," in *The Cognitive Neurosciences*, 4th ed., ed. M. S. Gazzaniga, 309–20 (MIT Press, 2009).

7. A. Clark, *Surfing Uncertainty: Prediction, Action, and the Embodied Mind* (Oxford University Press, 2015); A. Clark, "Whatever Next? Predictive Brains, Situated Agents, and the Future of Cognitive Science," *Behavioral and Brain Sciences* 36, no. 3 (2013): 181–204; K. Friston, "A Theory of Cortical Responses," *Philosophical Transactions of the Royal Society B: Biological Sciences* 360, no. 1456 (2005): 815–36; J. Hohwy, *The Predictive Mind* (Oxford University Press, 2013).

8. W. James, *Writings 1902–1910: The Varieties of Religious Experience, Pragmatism, A Pluralistic Universe, The Meaning of Truth, Some Problems of Philosophy, Essays*, vol. 2 (Library of America, 1988).

9. Quoted in J. L. Garfield (trans.), *The Fundamental Wisdom of the Middle Way: Nagarjuna's Mulamadhyamakakarika* (Oxford University Press, 1995).

10. T. N. Hạnh, *No Death, No Fear: Comforting Wisdom for Life* (Penguin, 2003).

CHAPTER THREE

1. L. Po, "Zazen on Ching-t'ing Mountain," in *Crossing the Yellow River: Three Hundred Poems from the Chinese*, trans. S. Hamill (Tiger Bark Press, 2013).

2. D. Hume, *A Treatise of Human Nature* (Oxford University Press, 2000).

3. D. C. Dennett, *Consciousness Explained* (Penguin UK, 1993).

4. Paracelsus, *De natura rerum*, trans. C. H. Sisson (Routledge, 2003).

5. W. R. Newman, *Promethean Ambitions: Alchemy and the Quest to Perfect Nature* (University of Chicago Press, 2019).

6. B. Franco, B. Marco, F. Roberto, and F. Francesca, "Point Zero: A Phenomenological Inquiry into the Subjective Physical Location of Consciousness," *Perceptual and Motor Skills* 107 (2008): 323–35.

7. M. E. Raichle, A. M. MacLeod, A. Z. Snyder, W. J. Powers, D. A. Gusnard, and G. L. Shulman, "A Default Mode of Brain Function," *Proceedings of the National Academy of Sciences* 98, no. 2 (2001): 676–82.

8. M. Woollacott and A. Shumway-Cook, "The Mystical Experience and Its Neural Correlates," *Journal of Near-Death Studies* 38, no. 1 (2020): 3–25.

9. T. Metzinger, *Being No One: The Self-Model Theory of Subjectivity* (MIT Press, 2004).

10. R. L. Carhart-Harris and K. J. Friston, "REBUS and the Anarchic Brain: Toward a Unified Model of the Brain Action of Psychedelics," *Pharmacological Reviews* 71, no. 3 (2019): 316–44.

11. S. Chandaria, "The Bayesian Brain and Meditation," Oxford, UK, November 10, 2022, video, 1:42:07, https://hedonia.kringelbach.org/2023/01/01/the-bayesian-brain-and-meditation/.

12. Rumi, *The Book of Love: Poems of Ecstasy and Longing*, trans. C. Barks (HarperCollins, 2003).

13. T. Metzinger, "Minimal Phenomenal Experience: Meditation, Tonic Alertness, and the Phenomenology of 'Pure' Consciousness," *Philosophy and the Mind Sciences* 1, no. 1 (2020): 1–44.

CHAPTER FOUR

1. Lao Tzu, *Tao te ching* (Wordsworth, 1997), vii–xix.

2. F. Jackson, "Epiphenomenal Qualia," in *Artificial Intelligence and Cognitive Science*, vol. 3, *Consciousness and Emotion in Cognitive Science*, ed. A. Clark and J. Torbido, 197–206 (Routledge, 1998); F. Jackson, "What Mary Didn't Know," *Journal of Philosophy* 83, no. 5 (1986): 291–95.

3. R. Wright, *Why Buddhism Is True: The Science and Philosophy of Meditation and Enlightenment* (Simon and Schuster, 2017).

4. A. Huxley, *The Perennial Philosophy* (Harper and Brothers, 1945).

5. H. S. Rahi, *Sri Guru Granth Sahib Discovered: A Reference Book of Quotations from the Adi Granth* (Motilal Banarsidass, 1999).

6. B. De Spinoza, *A Spinoza Reader: The Ethics and Other Works* (Princeton University Press, 1994).

7. "Ban of Baruch Spinoza, Amsterdam, 27 July 1656," Wikimedia Commons, https://commons.wikimedia.org/wiki/File:Ban_of_Baruch_Spinoza,_Amsterdam,_27_July_1656,_6_Av_5416.jpg.

8. J. Maffie, *Aztec Philosophy: Understanding a World in Motion* (University Press of Colorado, 2014).

9. T. M. Robinson, ed., *Heraclitus: Fragments*, vol. 2 (University of Toronto Press, 1987).

10. J. Liu and D. L. Berger, eds., *Nothingness in Asian philosophy* (New York: Routledge, 2014); Nishida K., *Nishida Kitarō zenshū* [*Collected Works of Nishida Kitarō*] (Iwanami, 2000).

11. J. P. Sartre, *Being and Nothingness: An Essay on Phenomenological Ontology*, Éditions Gallimard (Philosophical Library, 1943).

12. R. Pine, *The Heart Sutra* (Catapult, 2005).

13. E. P. Tyron, "Is the Universe a Vacuum Fluctuation?" *Nature* 246, no. 5433 (1973): 396–97.

CHAPTER FIVE

1. A. Aguirre, *Cosmological Koans: A Journey to the Heart of Physical Reality* (W. W. Norton, 2019).

2. H. Everett III, "'Relative State' Formulation of Quantum Mechanics," *Reviews of Modern Physics* 29, no. 3 (1957): 454.

3. A. Aspect, "Closing the Door on Einstein and Bohr's Quantum Debate," *Physics* 8 (2015): 123; D. Bouwmeester, J. W. Pan, K. Mattle, M. Eibl, H. Weinfurter, and A. Zeilinger, "Experimental Quantum Teleportation," *Nature* 390, no. 6660 (1997): 575–79; S. J. Freedman and J. F. Clauser, "Experimental Test of Local Hidden-Variable Theories," *Physical Review Letters* 28, no. 14 (1972): 938.

4. F. Capra, *The Tao of Physics: An Exploration of the Parallels between Modern Physics and Eastern Mysticism* (Shambhala, 1975).

5. C. Rovelli, *Helgoland* (Flammarion, 2023).

6. J. A. Wheeler, "Information, Physics, Quantum: The Search for Links," in *Complexity, Entropy, and the Physics of Information*, ed. W. H. Zurek (Addison-Wesley, 1990).

7. Capra, *Tao of Physics*.

8. J. S. Bell, "On the Einstein Podolsky Rosen Paradox," *Physics Physique Fizika* 1, no. 3 (1964): 195.

9. I. Kant, "Critique of Pure Reason (1781)," in *Modern Classical Philosophers: Selections Illustrating Modern Philosophy from Bruno to Spencer*, comp. B. Rand, 370–456 (Houghton Mifflin, 1908).

CHAPTER SIX

1. M. Planck, interview in *Observer*, January 25, 1931, 17, col. 3, https://en.wikiquote.org/wiki/Max_Planck.

2. G. Berkeley, *A Treatise Concerning the Principles of Human Knowledge* (A. Rhames for J. Pepyat, 1710).

3. B. Kastrup, *The Idea of the World: A Multi-Disciplinary Argument for the Mental Nature of Reality* (John Hunt, 2019).

4. S. Nishimoto, A. T. Vu, T. Naselaris, Y. Benjamini, B. Yu, and J. L. Gallant, "Reconstructing Visual Experiences from Brain Activity Evoked by Natural Movies," *Current Biology* 21, no. 19 (2011): 1641–46.

5. B. Russell, *My Philosophical Development* (Routledge, 1995).

6. G. Galilei, "The Assayer," in *Discoveries and Opinions of Galileo*, trans. S. Drake (Anchor, 1957).

7. P. Goff, *Galileo's Error: Foundations for a New Science of Consciousness* (Vintage, 2019).

8. A. N. Whitehead, *Science and Philosophy* (Open Road Media, 2014).

9. M. McGin, *The Routledge Guidebook to Wittgenstein's Philosophical Investigations* (Routledge, 2013).

10. A. N. Whitehead, "Process and Reality: An Essay in Cosmology," in *Gifford Lectures Delivered in the University of Edinburgh during the Session 1927–1928* (Macmillan, 1929).

11. A. Osborne, ed., *The Collected Works of Ramana Maharshi* (Weiser Books, 1997).

12. R. T. Griffith (trans.), *The Rig Veda*, vol. 1 (Library of Alexandria, 2013).

13. B. C. Muraresku, *The Immortality Key: The Secret History of the Religion with No Name* (St. Martin's Press, 2020).

14. C. Timmermann, H. Kettner, C. Letheby, L. Roseman, F. E. Rosas, and R. L. Carhart-Harris, "Psychedelics Alter Metaphysical Beliefs," *Scientific Reports* 11, no. 1 (2021): 22166.

15. D. Lingpa, *Buddhahood without Meditation: A Visionary Account Known as Refining Apparent Phenomena*, trans. R. Barron (Padma, 1994).

CHAPTER SEVEN

1. R. Burns, "To a Mouse, on Turning Her Up In Her Nest with the Plough," in *The Poetry of Robert Burns*, vol. 1, ed. W. E. Henley and T. F. Henderson, 152–54 (Cornell University Library, 2009).

2. B. Russell, *Russell on Ethics: Selections from the Writings of Bertrand Russell*, ed. C. R. Pigden (Routledge, 2013).

3. P. Low, *The Cambridge Declaration on Consciousness*, ed. J. Panksepp, D. Reiss, D. Edelman, B. Van Swinderen, P. Low, and C. Koch, July 7, 2012, https://fcmconference.org/img/V9_Cambridge_Declaration_on_Consciousness.pdf.

4. D. Abram, *The Spell of the Sensuous: Perception and Language in a More-than-Human World* (Vintage, 2012).

5. J. Hickel, *Less Is More: How Degrowth Will Save the World* (Random House, 2020).

6. C. Merchant, *The Death of Nature: Women, Ecology and the Scientific Revolution* (Harper and Row, 1980).

7. Abram, *Spell of the Sensuous*.

8. A. Korzybski, *Science and Sanity: An Introduction to Non-Aristotelian Systems and General Semantics* (International Non-Aristotelian Library, 1933).

9. Genesis 1:26 (New King James Version).

10. R. Valantasis, *The Gospel of Thomas* (Routledge, 2008).

11. G. Pico della Mirandola, *Opera omnia* (1557; Olms, 1969).

12. J. Muir, "My First Summer in the Sierra," in *British Politics and the Environment in the Long Nineteenth Century*, ed. P. Hough, 291–96 (Routledge, 2023).

13. E. Haeckel, *Monism as Connecting Religion and Science: The Confession of Faith of a Man of Science* (A. and C. Black, 1894).

14. Haeckel, *Monism as Connecting Religion*.

15. R. Carson, *Silent Spring* (Houghton Mifflin, 1962).

16. A. Næss, "The Shallow and the Deep: Long-Range Ecology Movement: A Summary," *Inquiry* 16 (1973): 95–100.

17. A. Naess, "Spinoza and Ecology," in *Speculum Spinozanum, 1677–1977*, ed. S. Hessing, 418–25 (Routledge, 2019).

18. T. Lyons and R. L. Carhart-Harris, "Increased Nature Relatedness and Decreased Authoritarian Political Views after Psilocybin for Treatment-Resistant Depression," *Journal of Psychopharmacology* 32, no. 7 (2018): 811–19.

19. R. Watts, C. Day, J. Krzanowski, D. Nutt, and R. Carhart-Harris, "Patients' Accounts of Increased 'Connectedness' and 'Acceptance' after Psilocybin for Treatment-Resistant Depression," *Journal of Humanistic Psychology* 57, no. 5 (2017): 520–64.

20. H. Kettner, S. Gandy, E. C. Haijen, and R. L. Carhart-Harris, "From Egoism to Ecoism: Psychedelics Increase Nature Relatedness in a State-Mediated and Context-Dependent Manner," *International Journal of Environmental Research and Public Health* 16, no. 24 (2019): 5147.

21. R. M. Doyle, *Darwin's Pharmacy: Sex, Plants, and the Evolution of the Noosphere* (University of Washington Press, 2011).

22. A. Einstein, in A. Einstein, J. Dewey, J. Jeans, H. G. Wells, T. Dreiser, H. L. Mencken, J. Truslow et al., *Living Philosophies* (AMS Press, 1931).

CHAPTER EIGHT

1. E. Dickinson, *The Collected Poems of Emily Dickinson* (First Avenue Editions, 2016).

2. C. McGinn, "Can We Solve the Mind-Body Problem?" *Mind* 98, no. 391 (1989): 349–66.

3. F. Crick, *The Astonishing Hypothesis: The Scientific Search for the Soul* (Simon and Schuster, 1994).

4. M. Glickstein and D. Whitteridge, "Tatsuji Inouye and the Mapping of the Visual Fields on the Human Cerebral Cortex," *Trends in Neurosciences* 10, no. 9 (1987): 350–53.

5. G. Riddoch, "Dissociation of Visual Perceptions Due to Occipital Injuries, with Especial Reference to Appreciation of Movement," *Brain* 40, no. 1 (1917): 15–57.

6. L. Weiskrantz, E. K. Warrington, M. D. Sanders, and J. Marshall, "Visual Capacity in the Hemianopic Field Following a Restricted Occipital Ablation," *Brain* 97, no. 1 (1974): 709–28; E. Pöppel, R. Held, and D. Frost, "Residual Visual Function after Brain Wounds Involving the Central Visual Pathways in Man," *Nature* 243, no. 5405 (1973): 295–96.

7. D. H. Hubel and T. N. Wiesel, "Receptive Fields, Binocular Interaction and Functional Architecture in the Cat's Visual Cortex," *Journal of Physiology* 160, no. 1 (1962): 106.

8. S. He, P. Cavanagh, and J. Intriligator, "Attentional Resolution and the Locus of Visual Awareness," *Nature* 383, no. 6598 (1996): 334–37.

9. A. Cowey and C. A. Heywood, "Cerebral Achromatopsia: Colour Blindness Despite Wavelength Processing," *Trends in Cognitive Sciences* 1, no. 4 (1997): 133–39.

10. S. Zeki, "Cerebral Akinetopsia (Visual Motion Blindness): A Review," *Brain* 114, no. 2 (1991): 811–24.

11. M. Boly, M. Massimini, N. Tsuchiya, B. R. Postle, C. Koch, and G. Tononi, "Are the Neural Correlates of Consciousness in the Front or in the Back of the

Cerebral Cortex? Clinical and Neuroimaging Evidence," *Journal of Neuroscience* 37, no. 40 (2017): 9603–13.

12. B. J. Baars, *A Cognitive Theory of Consciousness* (Cambridge University Press, 1993).

13. S. Frässle, J. Sommer, A. Jansen, M. Naber, and W. Einhäuser, "Binocular Rivalry: Frontal Activity Relates to Introspection and Action but Not to Perception," *Journal of Neuroscience* 34, no. 5 (2014): 1738–47.

14. A. D. Craig, "How Do You Feel—Now? The Anterior Insula and Human Awareness," *Nature Reviews Neuroscience* 10, no. 1 (2009): 59–70.

15. A. R. Damasio, *The Feeling of What Happens: Body and Emotion in the Making of Consciousness* (Houghton Mifflin Harcourt, 1999).

16. Craig, "How Do You Feel."

17. J. Parvizi and A. Damasio, "Consciousness and the Brainstem," *Cognition* 79, nos. 1–2 (2001): 135–60; M. Solms, *The Hidden Spring: A Journey to the Source of Consciousness* (Profile Books, 2021).

CHAPTER NINE

1. R. Penrose, *The Emperor's New Mind: Concerning Computers, Minds, and the Laws of Physics* (Oxford University Press, 1999).

2. J. McFadden, *Quantum Evolution* (HarperCollins, 2000); J. McFadden, "Synchronous Firing and Its Influence on the Brain's Electromagnetic Field," *Journal of Consciousness Studies* 9, no. 4 (2002): 23–50; S. Pockett, *The Nature of Consciousness: A Hypothesis* (IUniverse, 2000).

3. R. Sperry, "Consciousness, Personal Identity and the Divided Brain," *Neuropsychologia* 22, no. 6 (1984): 661–73.

4. M. S. Gazzaniga, "Principles of Human Brain Organization Derived from Split-Brain Studies," *Neuron* 14, no. 2 (1995): 217–28.

5. S. Hameroff and R. Penrose, "Orchestrated Reduction of Quantum Coherence in Brain Microtubules: A Model for Consciousness," *Mathematics and Computers in Simulation* 40, nos. 3–4 (1996): 453–80.

6. G. Tononi, "An Information Integration Theory of Consciousness," *BMC Neuroscience* 5 (2004): 1–22.

7. H. Putnam, "Minds and Machines," in *Dimensions of Minds*, ed. Sidney Hook (New York University Press, 1960).

8. J. R. Searle, *The Rediscovery of the Mind* (MIT Press, 1992).

CHAPTER TEN

1. J. Joyce, *Dubliners* (Oxford University Press, 2008).

2. J. J. Gibson, *The Senses Considered as Perceptual Systems* (Houghton Mifflin, 1966).

3. H. R. Maturana and F. J. Varela, *Autopoiesis and Cognition: The Realization of the Living*, vol. 42 (Springer Science and Business Media, 1980).

4. F. J. Varela, E. Thompson, and E. Rosch, *The Embodied Mind: Cognitive Science and Human Experience*, rev. ed. (MIT Press, 2017).

5. A. Noë, *Action in Perception* (MIT Press, 2004); M. Sheets-Johnstone, *The Primacy of Movement* (John Benjamins, 2004).

6. R. Hanna and M. Maiese, *Embodied Minds in Action* (Oxford University Press, 2009).

7. H. von Helmholtz, *Helmholtz's Treatise on Physiological Optics*, vol. 3 (Optical Society of America, 1925).

8. P. Dayan, G. E. Hinton, R. M. Neal, and R. S. Zemel, "The Helmholtz Machine," *Neural Computation* 7, no. 5 (1995): 889–904.

9. R. P. Rao and D. H. Ballard, "Predictive Coding in the Visual Cortex: A Functional Interpretation of Some Extra-Classical Receptive-Field Effects," *Nature Neuroscience* 2, no. 1 (1999): 79–87.

10. I. Kant, "Critique of Pure Reason (1781)," in *Modern Classical Philosophers*, Cambridge, comp. B. Rand, 370–456 (Houghton Mifflin, 1908).

11. K. Friston, "Consciousness and Hierarchical Inference," *Neuropsychoanalysis* 15, no. 1 (2013): 38–42.

12. A. Tversky and D. Kahneman, "Judgment under Uncertainty: Heuristics and Biases: Biases in Judgments Reveal Some Heuristics of Thinking under Uncertainty," *Science* 185, no. 4157 (1974): 1124–31.

13. K. Friston, J. Kilner, and L. Harrison, "A Free Energy Principle for the Brain," *Journal of Physiology—Paris* 100, nos. 1–3 (2006): 70–87.

14. T. Parr, G. Pezzulo, and K. J. Friston, *Active Inference: The Free Energy Principle in Mind, Brain, and Behavior* (MIT Press, 2022).

15. A. Clark, "Consciousness as Generative Entanglement," *Journal of Philosophy* 116, no. 12 (2019): 645–62; D. Rudrauf, G. Sergeant-Perthuis, Y. Tisserand, G. Poloudenny, K. Williford, and M. A. Amorim, "The Projective Consciousness Model: Projective Geometry at the Core of Consciousness and the Integration of Perception, Imagination, Motivation, Emotion, Social Cognition and Action," *Brain Sciences* 13, no. 10 (2023): 1435; A. Safron, "An Integrated World Modeling Theory (IWMT) of Consciousness: Combining Integrated Information and Global Neuronal Workspace Theories with the Free Energy

Principle and Active Inference Framework: Toward Solving the Hard Problem and Characterizing Agentic Causation," *Frontiers in Artificial Intelligence* 3 (2020): 30; A. K. Seth and M. Tsakiris, "Being a Beast Machine: The Somatic Basis of Selfhood," *Trends in Cognitive Sciences* 22, no. 11 (2018): 969–81.

16. M. Solms, *The Hidden Spring: A Journey to the Source of Consciousness* (Profile Books, 2021).

17. W. Wiese, "The Science of Consciousness Does Not Need Another Theory, It Needs a Minimal Unifying Model," *Neuroscience of Consciousness* 2020, no. 1 (2020): niaa013.

18. M. J. Ramstead, M. Albarracin, A. Kiefer, B. Klein, C. Fields, K. Friston, and A. Safron, "The Inner Screen Model of Consciousness: Applying the Free Energy Principle Directly to the Study of Conscious Experience," *arXiv* (2023): 2305.02205.

19. A. Seth, *Being You: A New Science of Consciousness* (Penguin, 2021).

20. J. Hohwy and A. Seth, "Predictive Processing as a Systematic Basis for Identifying the Neural Correlates of Consciousness," *Philosophy and the Mind Sciences* 1, no. 2 (2020).

21. S. Rushdie, *The Satanic Verses: A Novel* (Picador, 1997).

22. T. W. Deacon, *Incomplete Nature: How Mind Emerged from Matter* (W. W. Norton, 2011); M. H. Bickhard, "The Interactivist Model," *Synthese* 166 (2009): 547–91.

23. A. Kolchinsky and D. H. Wolpert, "Semantic Information, Autonomous Agency and Non-Equilibrium Statistical Physics," *Interface Focus* 8, no. 6 (2018): 20180041.

CHAPTER ELEVEN

1. E. Schrödinger, *What Is Life? The Physical Aspect of the Living Cell* (Cambridge University Press, 1943).

2. E. O. Wilson, *Consilience: The Unity of Knowledge*, vol. 31 (Vintage, 1999).

3. B. De Spinoza, *A Spinoza Reader: The Ethics and Other Works* (Princeton University Press, 1994).

4. R. Wicks, "Arthur Schopenhauer," in *Stanford Encyclopedia of Philosophy*, ed. E. N. Zanta (Stanford University, Fall 2021).

5. H. Bergson, *Creative Evolution* (1911), trans. A. Mitchell (Dover, 1998).

6. Schrödinger, *What Is Life?*

7. I. Prigogine, "Dissipative Structures, Dynamics and Entropy," *International Journal of Quantum Chemistry* 9, no. S9 (1975): 443–56; I. Prigogine and R. Lefever,

"Theory of Dissipative Structures," in *Synergetics: Cooperative Phenomena in Multi-Component Systems*, ed. H. Haken, 124–35 (Vieweg+Teubner Verlag Wiesbaden, 1973).

8. D. Kondepudi, B. Kay, and J. Dixon, "End-Directed Evolution and the Emergence of Energy-Seeking Behavior in a Complex System," *Physical Review E* 91, no. 5 (2015): 050902.

9. G. E. Crooks, "Entropy Production Fluctuation Theorem and the Nonequilibrium Work Relation for Free Energy Differences," *Physical Review E* 60, no. 3 (1999): 2721.

10. J. L. England, "Dissipative Adaptation in Driven Self-Assembly," *Nature Nanotechnology* 10, no. 11 (2015): 919–23.

11. N. Wolchover, "A New Physics Theory of Life," *Quanta*, January 22, 2014, https://www.quantamagazine.org/a-new-thermodynamics-theory-of-the-origin-of-life-20140122/.

12. Quoted in L. Pasteur, "On Spontaneous Generation," address delivered at the Sorbonne Scientific Soirée, 1864.

13. M. P. Robertson and G. F. Joyce, "The Origins of the RNA World," *Cold Spring Harbor Perspectives in Biology* 4, no. 5 (2012): a003608.

14. D. Segré, D. Ben-Eli, D. W. Deamer, and D. Lancet, "The Lipid World," *Origins of Life and Evolution of the Biosphere* 31 (2001): 119–45.

15. S. A. Kauffman, "Cellular Homeostasis, Epigenesis and Replication in Randomly Aggregated Macromolecular Systems," *Journal of Cybernetics* 1, no. 1 (1971): 71–96.

16. S. L. Miller and H. C. Urey, "Organic Compound Synthesis on the Primitive Earth: Several Questions about the Origin of Life Have Been Answered, but Much Remains to Be Studied," *Science* 130, no. 3370 (1959): 245–51.

17. S. Iglesias-Groth, "A Search for Tryptophan in the Gas of the IC 348 Star Cluster of the Perseus Molecular Cloud," *Monthly Notices of the Royal Astronomical Society* 523, no. 2 (2023): 2876–86.

18. C. Darwin, letter to J. D. Hooker, February 1, 1871, Darwin Correspondence Project, https://www.darwinproject.ac.uk/letter/?docId=letters/DCP-LETT-7471.xml; A. I. Oparin, *The Origin of Life*, trans. S. Morgulis (1924; Dover, 2003).

19. R. F. Weiss, P. Lonsdale, J. E. Lupton, A. E. Bainbridge, and H. Craig, "Hydrothermal Plumes in the Galapagos Rift," *Nature* 267, no. 5612 (1977): 600–603.

20. D. S. Kelley, J. A. Karson, G. L. Fruh-Green, D. R. Yoerger, T. M. Shank, D. A. Butterfield, J. M. Hayes, et al., "A Serpentinite-Hosted Ecosystem: The Lost City Hydrothermal Field," *Science* 307, no. 5714 (2005): 1428–34.

21. E. Camprubi, S. F. Jordan, R. Vasiliadou, and N. Lane, "Iron Catalysis at the Origin of Life," *IUBMB Life* 69, no. 6 (2017): 373–81; N. Lane, *Transformer: The Deep Chemistry of Life and Death* (Profile Books, 2022).

22. Camprubi et al., "Iron Catalysis."

23. T. Dobzhansky, "Nothing in Biology Makes Sense Except in the Light of Evolution," *American Biology Teacher* 35, no. 3 (1973): 125–29.

24. R. Dawkins, *The Selfish Gene* (Oxford University Press, 2016).

25. J. O. Campbell and M. E. Price, "Universal Darwinism and the Origins of Order," in *Evolution, Development and Complexity: Multiscale Evolutionary Models of Complex Adaptive Systems*, ed. G. Y. Georgiev, J. M. Smart, C. L. Flores Martinez, and M. E. Price, 261–90 (Springer International, 2019); R. Dawkins, "Universal Darwinism," in *Evolution from Molecules to Man*, ed. D. S. Bendall, 403–25 (Cambridge University Press, 1983).

26. B. F. Skinner, *The Behavior of Organisms: An Experimental Analysis* (Appleton-Century-Crofts, 1938).

27. B. F. Skinner, "Selection by Consequences," *Behavioral and Brain Sciences* 7, no. 4 (1984): 477–81.

28. G. M. Edelman, "Group Selection and Phasic Re-entrant Signalling: A Theory of Higher Brain Function," in *The Mindful Brain*, ed. G. M. Edelman and V. B. Mountcastle, 51–100 (MIT Press, 1978).

29. D. Lewis, "Causation," *Journal of Philosophy* 70, no. 17 (1973): 556–67.

30. W. Charlton, ed., *Aristotle's Physics: Books I and II* (Oxford University Press, 1983).

31. D. Papineau, "The Causal Closure of the Physical and Naturalism," in *The Oxford Handbook of Philosophy of Mind*, ed. B. McLaughlin, A. Beckermann, and S. Walter, 53–65 (Oxford University Press, 2009).

32. A. Juarrero, "Causality as Constraint," in *Evolutionary Systems: Biological and Epistemological Perspectives on Selection and Self-Organization*, ed. G. Vijver, S. N. Salthe, and M. Delpos, 233–42 (Springer Netherlands, 1998); A. Juarrero, *Context Changes Everything: How Constraints Create Coherence* (MIT Press, 2023); A. Juarrero, *Dynamics in Action: Intentional Behavior as a Complex System* (MIT Press, 1999).

33. S. A. Kauffman, *A World beyond Physics: The Emergence and Evolution of Life* (Oxford University Press, 2019).

34. M. Montévil and M. Mossio, "Biological Organisation as Closure of Constraints," *Journal of Theoretical Biology* 372 (2015): 179–91.

35. E. Thompson, *Mind in Life: Biology, Phenomenology, and the Sciences of Mind* (Harvard University Press, 2010); F. Capra and P. L. Luisi, *The Systems View of Life: A Unifying Vision* (Cambridge University Press, 2014).

CHAPTER TWELVE

1. Hadewijch II, "You Who Want . . . ," in *Women in Praise of the Sacred*, trans. J. Hirshfield (HarperCollins, 1994).

2. C. E. Shannon, "A Mathematical Theory of Communication," *Bell System Technical Journal* 27, no. 3 (1948): 379–423.

3. K. Friston, "Life as We Know It," *Journal of the Royal Society Interface* 10, no. 86 (2013): 20130475.

4. J. E. Cooke, "The Living Mirror Theory of Consciousness," *Journal of Consciousness Studies* 27, nos. 9–10 (2020): 127–47.

5. Plato, *Theaetetus* (BoD—Books on Demand, 2019).

6. C. McGinn, "Consciousness as Knowingness," *Monist* 91, no. 2 (2008): 237–49.

7. T. Metzinger, "Minimal Phenomenal Experience: Meditation, Tonic Alertness, and the Phenomenology of 'Pure' Consciousness," *Philosophy and the Mind Sciences* 1, no. 1 (2020): 1–44.

8. H. James, "The Art of Fiction," in *The Writer's Art: By Those Who Have Practiced It*, ed. R. W. Brown, 210–31 (Harvard University Press, 1924).

CHAPTER THIRTEEN

1. P. H. Barrett, ed., *The Works of Charles Darwin*, vol. 27, *The Power of Movement in Plants* (Taylor and Francis, 2016).

2. E. Thompson, *Mind in Life: Biology, Phenomenology, and the Sciences of Mind* (Harvard University Press, 2010).

3. E. Thompson, "Could All Life Be Sentient?" *Journal of Consciousness Studies* 29, nos. 3–4 (2022): 229–65.

4. L. Margulis and D. Sagan, *What Is Life?* (University of California Press, 2000).

5. A. S. Reber, "Caterpillars and Consciousness," *Philosophical Psychology* 10, no. 4 (1997): 437–49; A. S. Reber, "Caterpillars, Consciousness and the Origins of Mind," *Animal Sentience* 1, no. 11 (2016): 1; A. S. Reber, *The First Minds: Caterpillars, 'Karyotes, and Consciousness* (Oxford University Press, 2018).

6. F. Baluška, S. Mancuso, and D. Volkmann, eds., *Communication in Plants: Neuronal Aspects of Plant Life* (Springer Verlag, 2006).

7. A. S. Reber, "Resolving the Hard Problem and Calling for a Small Miracle," *Animal Sentience* 1, no. 11 (2016): 9; Thompson, "Could All Life Be Sentient?"

8. T. W. Deacon, *Incomplete Nature: How Mind Emerged from Matter* (W. W. Norton, 2011).

9. M. H. Bickhard, "The Interactivist Model," *Synthese* 166 (2009): 547–91.

10. A. R. Damasio, *The Feeling of What Happens: Body and Emotion in the Making of Consciousness* (Houghton Mifflin Harcourt, 1999).

11. M. Solms, *The Hidden Spring: A Journey to the Source of Consciousness* (Profile Books, 2021).

12. R. Lanza and B. Berman, *Biocentrism: How Life and Consciousness Are the Keys to Understanding the True Nature of the Universe* (BenBella Books, 2010).

13. A. Fedorov, A. Lehto, and J. Klein, "Inhibition of Mitochondrial Respiration by General Anesthetic Drugs," *Naunyn-Schmiedeberg's Archives of Pharmacology* 396, no. 2 (2023): 375–81.

14. W. James, *The Principles of Psychology* (H. Holt, 1890).

15. Y. Akhtar and M. B. Isman, "Larval Exposure to Oviposition Deterrents Alters Subsequent Oviposition Behavior in Generalist, *Trichoplusia ni* and Specialist, *Plutella xylostella* Moths," *Journal of Chemical Ecology* 29 (2003): 1853–70.

16. F. Durant, J. Morokuma, C. Fields, K. Williams, D. S. Adams, and M. Levin, "Long-Term, Stochastic Editing of Regenerative Anatomy via Targeting Endogenous Bioelectric Gradients," *Biophysical Journal* 112, no. 10 (2017): 2231–43.

17. H. S. Jennings, *Behavior of the Lower Organisms* (1906; Columbia University Press, 1931).

18. J. P. Dexter, S. Prabakaran, and J. Gunawardena, "A Complex Hierarchy of Avoidance Behaviors in a Single-Cell Eukaryote," *Current Biology* 29, no. 24 (2019): 4323–29.

19. D. C. Wood, "Habituation in Stentor: Produced by Mechanoreceptor Channel Modification," *Journal of Neuroscience* 8, no. 7 (1988): 2254–58.

20. B. Gelber, "Investigations of the Behavior of *Paramecium aurelia*: I. Modification of Behavior after Training with Reinforcement," *Journal of Comparative and Physiological Psychology* 45, no. 1 (1952): 58.

21. S. J. Gershman, P. E. Balbi, C. R. Gallistel, and J. Gunawardena, "Reconsidering the Evidence for Learning in Single Cells," *Elife* 10 (2021): e61907.

22. T. Nakagaki, H. Yamada, and Á. Tóth, "Maze-Solving by an Amoeboid Organism," *Nature* 407, no. 6803 (2000): 470.

23. A. Tero, S. Takagi, T. Saigusa, K. Ito, D. P. Bebber, M. D. Fricker, K. Yumiki, R. Kobayashi, and T. Nakagaki, "Rules for Biologically Inspired Adaptive Network Design," *Science* 327, no. 5964 (2010): 439–42.

24. A. Adamatzky, *Physarum Machines: Computers from Slime Mould*, vol. 74 (World Scientific, 2010).

25. M. Aono, Y. Hirata, M. Hara, and K. Aihara, "Amoeba-Based Chaotic Neurocomputing: Combinatorial Optimization by Coupled Biological Oscillators," *New Generation Computing* 27 (2009): 129–57.

26. J. Gomez-Ramirez and R. Sanz, "What the *Escherichia coli* Tells Neurons about Learning," in *Integral Biomathics: Tracing the Road to Reality*, ed. P. L. Simeonov, L. S. Smith, and A. C. Ehresmann, 41–55 (Springer Science and Business Media, 2012).

27. Gomez-Ramirez and Sanz, "What the *Escherichia coli*," 49.

28. A. Allen, "Plant Blindness," *BioScience* 53, no. 10 (2003): 926.

29. V. Latzel and Z. Münzbergová, "Anticipatory Behavior of the Clonal Plant *Fragaria vesca*," *Frontiers in Plant Science* 9 (2018): 1847.

30. V. Raja, P. L. Silva, R. Holghoomi, and P. Calvo, "The Dynamics of Plant Mutation," *Scientific Reports* 10, no. 1 (2020): 19465.

31. H. Shibaoka and T. Yamaki, "Studies on the Growth Movement of Sunflower Plant," *Scientific Papers of the College of General Education University of Tokyo* 9 (1959): 105–26.

32. S. A. Dudley and A. L. File, "Kin Recognition in an Annual Plant," *Biology Letters* 3, no. 4 (2007): 435–38.

33. P. Wohlleben, *The Hidden Life of Trees: What They Feel, How They Communicate—Discoveries from a Secret World*, vol. 1 (Greystone Books, 2016).

34. S. W. Simard, D. A. Perry, M. D. Jones, D. D. Myrold, D. M. Durall, and R. Molina, "Net Transfer of Carbon between Tree Species with Shared Ectomycorrhizal Fungi," *Nature* 388 (1997): 579–82.

35. S. K. Hewitt, D. S. Foster, P. S., Dyer, and S. V. Avery, "Phenotypic Heterogeneity in Fungi: Importance and Methodology," *Fungal Biology Reviews* 30, no. 4 (2016): 176–84; N. P. Money, "Hyphal and Mycelial Consciousness: The Concept of the Fungal Mind," *Fungal Biology* 125, no. 4 (2021): 257–59.

36. S. A. Ramesh, S. D. Tyerman, M. Gilliham, and B. Xu, "γ-Aminobutyric Acid (GABA) Signalling in Plants," *Cellular and Molecular Life Sciences* 74 (2017): 1577–1603.

37. E. D. Brenner, R. Stahlberg, S. Mancuso, J. Vivanco, F. Baluška, and E. Van Volkenburgh, "Plant Neurobiology: An Integrated View of Plant Signaling," *Trends in Plant Science* 11, no. 8 (2006): 413–19.

38. M. Toyota, D. Spencer, S. Sawai-Toyota, W. Jiaqi, T. Zhang, A. J. Koo, G. A. Howe, and S. Gilroy, "Glutamate Triggers Long-Distance, Calcium-Based Plant Defense Signaling," *Science* 361, no. 6407 (2018): 1112–15.

39. H. M. Lam, J. Chiu, M. H. Hsieh, L. Meisel, I. C. Oliveira, M. Shin, and G. Coruzzi, "Glutamate-Receptor Genes in Plants," *Nature* 396, no. 6707 (1998): 125–26.

40. P. Calvo and K. Friston, "Predicting Green: Really Radical (Plant) Predictive Processing," *Journal of the Royal Society Interface* 14, no. 131 (2017): 20170096.

41. F. Baluška, W. B. Miller Jr., and A. S. Reber, "Biomolecular Basis of Cellular Consciousness via Subcellular Nanobrains," *International Journal of Molecular Sciences* 22, no. 5 (2021): 2545.

CHAPTER FOURTEEN

1. I. De Ribera-Martin, "Aristotle: *De Anima* (*On the Soul*)," *Review of Metaphysics* 72, no. 3 (2019): 587–89.

2. F. Crick and C. Koch, "Towards a Neurobiological Theory of Consciousness," *Seminars in Neuroscience* 2 (1990): 263–75.

3. D. J. Chalmers, *The Conscious Mind: In Search of a Fundamental Theory* (Oxford Paperbacks, 1997).

4. P. Godfrey-Smith, *Metazoa: Animal Life and the Birth of the Mind* (Farrar, Straus, and Giroux, 2020).

5. K. J. Friston, W. Wiese, and J. A. Hobson, "Sentience and the Origins of Consciousness: From Cartesian Duality to Markovian Monism," *Entropy* 22, no. 5 (2020): 516.

6. K. Friston, "Am I Self-Conscious? (Or Does Self-Organization Entail Self-Consciousness?)," *Frontiers in Psychology* 9 (2018): 579.

7. T. Metzinger, "Minimal Phenomenal Experience: Meditation, Tonic Alertness, and the Phenomenology of 'Pure' Consciousness," *Philosophy and the Mind Sciences* 1, no. 1 (2020): 1–44.

8. E. Thompson, *Waking, Dreaming, Being: Self and Consciousness in Neuroscience, Meditation, and Philosophy* (Columbia University Press, 2014).

9. W. James, *The Principles of Psychology*, vol. 1 (Cosimo, 2007).

10. G. M. Edelman, "Group Selection and Phasic Re-entrant Signalling: A Theory of Higher Brain Function," in *The Mindful Brain*, ed. G. M. Edelman and V. B. Mountcastle, 51–100 (MIT Press, 1978).

11. Crick and Koch, "Towards a Neurobiological Theory."

12. A. Juarrero, "Causality as Constraint," in *Evolutionary Systems: Biological and Epistemological Perspectives on Selection and Self-Organization*, ed. G. Vijver, S. N. Salthe, and M. Delpos, 233–42 (Springer Netherlands, 1998); A. Juarrero,

Context Changes Everything: How Constraints Create Coherence (MIT Press, 2023); A. Juarrero, *Dynamics in Action: Intentional Behavior as a Complex System* (MIT Press, 1999).

13. G. Tononi, "An Information Integration Theory of Consciousness," *BMC Neuroscience* 5 (2004): 1–22.

14. S. Chandaria, "A Critique of the Integrated Information Theory of Consciousness," April 2020, YouTube video, 1:23:53, https://youtu.be/yrTiVX vWYE0?si=g2jslts6dyHJEeZq; J. E. Cooke, "What Is Consciousness? Integrated Information vs. Inference," *Entropy* 23, no. 8 (2021): 1032; A. Safron, "An Integrated World Modeling Theory (IWMT) of Consciousness: Combining Integrated Information and Global Neuronal Workspace Theories with the Free Energy Principle and Active Inference Framework; Toward Solving the Hard Problem and Characterizing Agentic Causation," *Frontiers in Artificial Intelligence* 3, no. 30 (2020).

15. Juarrero, *Context Changes Everything*.

16. V. Woolf, *The Common Reader: First Series* (1925; DigiCat, 2023).

17. J. G. Yoder, "Christiaan Huygens: Book on the Pendulum Clock (1673)," in *Landmark Writings in Western Mathematics 1640–1940*, ed. I. Grattan-Guinness, R. Cooke, L. Corry, P. Crépel, and N. Guicciardini, 33–45 (Elsevier Science, 2005).

18. K. Friston, "Life as We Know It," *Journal of the Royal Society Interface* 10, no. 86 (2013): 20130475.

19. "The New York Declaration on Animal Consciousness," New York University, April 19, 2024, https://sites.google.com/nyu.edu/nydeclaration/declar ation.

20. V. Woolf, *The Waves* (Feedbooks, 1931).

21. W. Whitman, *Walt Whitman's "Song of Myself": A Sourcebook and Critical Edition* (Psychology Press, 2005).

22. E. Pöppel, R. Held, and D. Frost, "Residual Visual Function after Brain Wounds Involving the Central Visual Pathways in Man," *Nature* 243, no. 5405 (1973): 295–96; L. Weiskrantz, E. K. Warrington, M. D. Sanders, and J. Marshall, "Visual Capacity in the Hemianopic Field Following a Restricted Occipital Ablation," *Brain* 97, no. 1 (1974): 709–28.

23. W. Whitman, *I Sing the Body Electric* (Phoenix, 1996).

24. Aristotle, *De Anima*.

25. D. K. Osbon, ed., *A Joseph Campbell Companion: Reflections on the Art of Living* (Harper Perennial, 1995).

EPILOGUE

1. J. G. Neihardt, *Black Elk Speaks: The Complete Edition* (University of Nebraska Press, 2014).

2. D. Whyte, *Essentials* (Canongate Books, 2022).

3. C. Darwin, *On the Origin of Species* (Penguin Classics, 2009).

4. K. P. Kropotkin, *Mutual Aid: A Factor of Evolution* (Black Rose Books, 2021).

5. A. Nin, *Seduction of the Minotaur* (Swallow Press, 1972).

6. M. Wolynn, *It Didn't Start with You: How Inherited Family Trauma Shapes Who We Are and How to End the Cycle* (Penguin, 2017).

7. S. Y. Appiah-Marfo, "Cultural Responses to Collective Trauma in Different Societies Explains Aspects of Their Identity, in *Comparative Criminology across Western and African Perspectives*, ed. S. P. Sungi and N. Ouassini, 137–58 (IGI Global, 2022).

8. R. L. Carhart-Harris and K. J. Friston, "REBUS and the Anarchic Brain: Toward a Unified Model of the Brain Action of Psychedelics," *Pharmacological Reviews* 71, no. 3 (2019): 316–44.

9. T. N. Hạnh, *Awakening of the Heart: Essential Buddhist Sutras and Commentaries* (Parallax Press, 2011).

10. E. Becker, *The Denial of Death* (Simon and Schuster, 1997).